TIME

100 New Scientific Discoveries

Fascinating, Unbelievable, and Mind-Expanding Stories

MAXIMILIEN BRICE/CERN

THE UNIVERSE MACHINE *Peering down just one small stretch of the Large Hadron Collider, Europe's new—and massive—particle accelerator*

TIME

MANAGING EDITOR Richard Stengel
DESIGN DIRECTOR D.W. Pine
DIRECTOR OF PHOTOGRAPHY Kira Pollack

100 New Scientific Discoveries

Fascinating, Unbelievable, and Mind-Expanding Stories

EDITOR Jeffrey Kluger
DESIGNER Sharon Okamoto
PHOTO EDITOR Dot McMahon
EDITORIAL PRODUCTION Lionel P. Vargas

TIME HOME ENTERTAINMENT
PUBLISHER Richard Fraiman
GENERAL MANAGER Steven Sandonato
EXECUTIVE DIRECTOR, MARKETING SERVICES Carol Pittard
EXECUTIVE DIRECTOR, RETAIL AND SPECIAL SALES Tom Mifsud
EXECUTIVE DIRECTOR, NEW PRODUCT DEVELOPMENT Peter Harper
DIRECTOR, BOOKAZINE DEVELOPMENT AND MARKETING Laura Adam
PUBLISHING DIRECTOR, BRAND MARKETING Joy Butts
ASSISTANT GENERAL COUNSEL Helen Wan
BOOK PRODUCTION MANAGER Suzanne Janso
DESIGN AND PREPRESS MANAGER Anne-Michelle Gallero
BRAND MANAGER Michela Wilde

SPECIAL THANKS TO:
Christine Austin, Jeremy Biloon, Glenn Buonocore, Malati Chavali, Jim Childs, Susan Chodakiewicz, Rose Cirrincione, Jacqueline Fitzgerald, Christine Font, Carrie Frazier, Lauren Hall, Malena Jones, Mona Li, Robert Marasco, Kimberly Marshall, Amy Migliaccio, Nina Mistry, Brooke Reger, Dave Rozzelle, Ilene Schreider, Adriana Tierno, Alex Voznesenskiy, Vanessa Wu, Time Imaging

Published by TIME Books, an imprint of Time Home Entertainment Inc.
135 West 50th Street, New York, N.Y. 10020

ISBN 10: 1-60320-172-6
ISBN 13: 978-1-60320-172-8
Library of Congress Control Number: 2010941158

We welcome your comments and suggestions about TIME Books. Please write to us at: TIME Books, Attention: Book Editors, P.O. Box 11016, Des Moines, IA 50336-1016

If you would like to order any of our hardcover Collector's Edition books, please call us at 1-800-327-6388, Monday through Friday, 7 a.m. to 8 p.m., or Saturday, 7 a.m. to 6 p.m. Central Time

MICK ELLISON

RORSCHACH DINO *A single specimen preserved between two slabs, this is one of the earliest-known feathered dinosaurs, with three kinds of plume.*

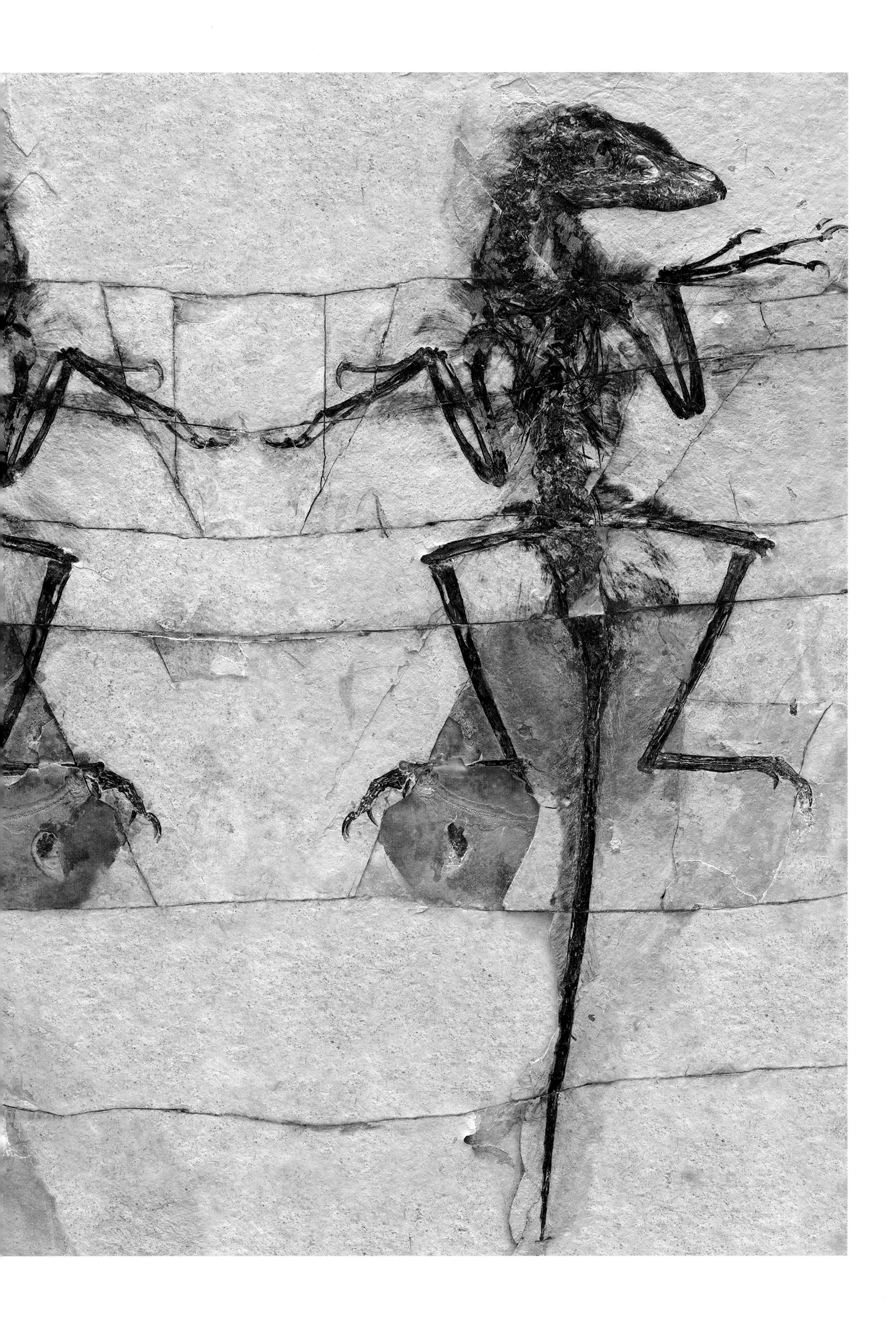

Contents

SKY EYES *The thin, dry air of Chile's Atacama Desert is a perfect spot for radio telescopes to observe the starry, moonlit sky.*

JOSÉ FRANCISCO SALGADO/ESO

The Grand Typhoon That Is Science

By Jeffrey Kluger

NOBODY CONVENED A PRESS CONFERENCE ON THAT UNREMARKED-UPON day in 1869 when Russian chemistry professor Dmitri Mendeleev developed the first workable periodic table of the elements. Nobody printed a banner headline back in 1610 when Galileo Galilei turned his telescope toward Jupiter and discovered that the planet was circled by four large moons. Like all the other scientists of centuries past, Galileo and Mendeleev just went about their work, trusting that the slow seep of knowledge would eventually carry the wisdom in their notebooks out to the wider world.

Things have changed a lot since then, and in most ways that's good. Science has become democratized—if not in the practice of it, at least in discussions about it. Governments fund labs and insist that the discoveries made there are revealed to the taxpayers who foot the bill. Benefactors endow universities and demand that the good works and surprise findings those knowledge factories produce are widely known. Wire services and websites curate the stream of announcements pouring from laboratories every day. News stories report this or that treatment or cure for this or that disease or condition—a breakthrough that's relevant only until it's upended the next day or lost entirely in the storm of other reports demanding equal attention. Galileo elegantly named the four Jovian moons he discovered Io, Europa, Ganymede, and Callisto—and schoolchildren eventually learned what to call them. Astronomers have since discovered 59 more. Could you name Kore or Mneme—or the awful-sounding S/2003 J 23? Could anyone?

The bacchanal of science news is too huge for even the most knowledge-hungry to sample and digest completely. That leads not only to overload, but to the risk of non-science masquerading as the real thing. Have you heard the news? The moon landings were faked! Vaccines cause autism! Climate change is a hoax! It's hooey, the lot of it, but the more information there is out there, the more rubbish that can slip by.

No single book can make all of this make sense. But TIME's team of science reporters can help, surveying the great disciplines and subdisciplines—medicine, chemistry, space, physics, technology, zoology, and more—for the dispatches that matter, that change things, that move the knowledge needle in ways that can affect us all. Scientists in all those fields will keep investigating, keep discovering—and TIME will keep reporting. The news may always change, but our commitment to staying atop it won't.

DAN SAELINGER/CORBIS

A.C. VOLTS
ON
OFF
ON
OFF

The Cosmos

KIDNAPPED COMETS ▪ HANGING OUT IN THE COSMOS ▪ A SPACECRAFT'S RETURN ENGAGEMENT
LIFE ON TITAN ▪ SUPERNOVA SECRETS ▪ ARSENIC-EATING EXTRATERRESTRIAL
SPOTTING INVISIBLE STARS ▪ JAPAN'S SPACE PUSH ▪ DARK MATTER AND BRIGHT GALAXIES

The New Hunt for Other Earths

Astronomers are discovering that the Milky Way is full of planets. The real prize now is finding the ones that can support life.

By Michael D. Lemonick

"There are 400 billion stars out there, just in our galaxy alone," Jodie Foster told an awestruck Matthew McConaughey in the movie *Contact.* "If only one out of a million have planets ... and if just one out of a million of those had life ... and if just one out of a million of those had intelligent life ... there would be literally millions of civilizations out there." Well, not quite. When you actually do the math with all those one-in-a-millions, the odds would be against there being any civilizations out there at all. But the basic reasoning is something astronomers have used for decades. It's an awfully big galaxy, and if just a fraction of the stars have planets on which life might take hold, the Milky Way could be teeming with biology.

Even as recently as 1997, when *Contact* premiered, scientists had no clue how many stars did have planets. There was ours, of course, plus a tiny handful of uninhabitable worlds orbiting nearby stars, which astronomers had finally been able to spot, after decades of trying, just a couple of years earlier. But the overall number was still a big question mark. Planets might be even rarer than Jodie Foster's character imagined—or they might be everywhere. That question mark is rapidly being rubbed out. Over the past 15 years or so, astronomers, pushing their telescopes to the limit, have boosted the trophy list from less than a half-dozen alien worlds to more than 500.

Now, thanks to an extraordinary space telescope called Kepler, that number is about to spiral dramatically upward. In February 2011, Kepler scientists announced that their spacecraft had spotted more than 1,200 new candidate planets. Most still have to be confirmed, and some will turn out to be false alarms—but probably, say the experts, no

FROM TOP: ESA; JAXA/AFP/GETTY IMAGES/NEWSCOM

SUPEREARTH *Gliese 581c is about five times the mass of Earth and orbits a dim, cool star.*

Goldilocks Worlds: Where Things Are Just Right for Life

Too Close
The heat from a star can boil off water from planets that venture too close and can warm their surfaces to deadly temperatures. Dry, airless Mercury and hothouse Venus illustrate the perils of proximity.

Just Right
Earth exists in the habitable zone, where liquid water can be present in abundance. Life as we know it can't exist without water. Life as we haven't imagined it is, admittedly, more of a riddle.

Too Distant
Space is a cold place, and you don't have to edge too far from your home star before water freezes solid. Atmosphere retains heat, and Mars might be a thriving world if it had held on to more of its air.

Three Ways to Spot Planets

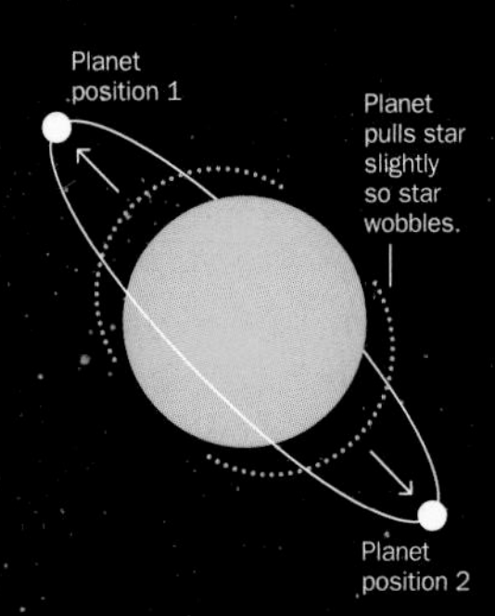

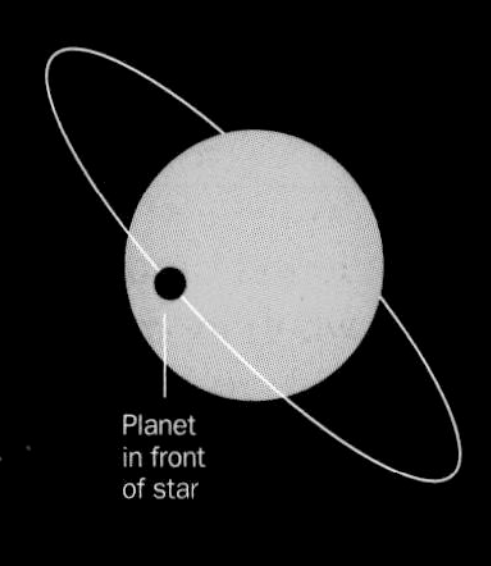

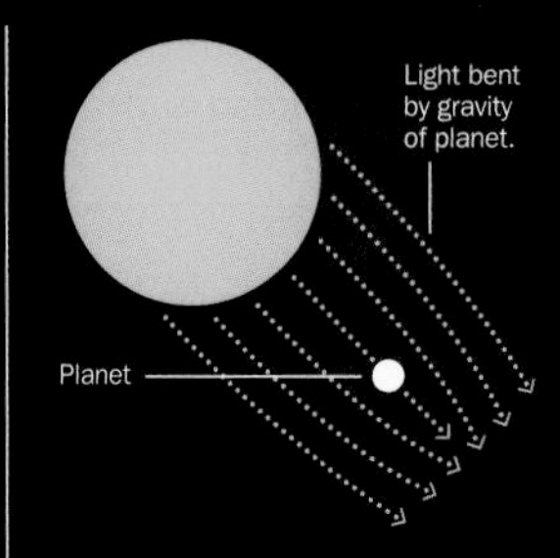

Wobble
As a planet orbits, it gravitationally tugs its parent star this way and that. By measuring this motion, scientists can verify a planet's existence and infer its mass.

Transit
Light from even the brightest star is slightly dimmed as an orbiting planet passes in front of it. The degree of dimming indicates the size of that planet.

Gravitational Microlensing
Gravity bends light, and a planet may thus distort the image of its star. This reveals the existence of a planet but little about its mass or diameter.

The New Kids on the Block

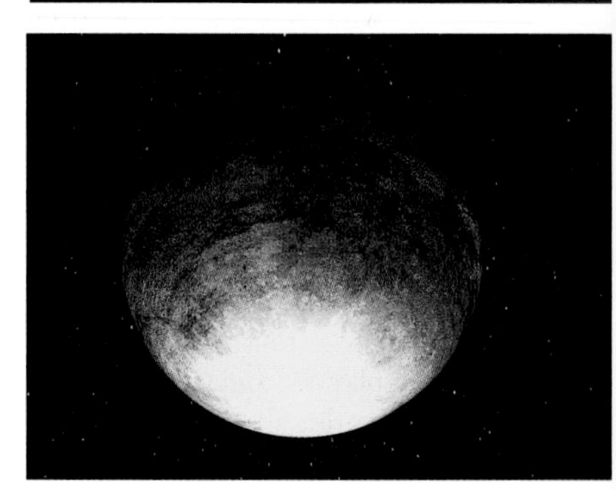

KEPLER-10B *About 4.6 times Earth's mass, it is the first rocky exoplanet known.*

HIP 13044 B *This Jupiter-like world was captured from another galaxy.*

GLIESE 581G *One of a family of at least four planets orbiting the red star Gliese*

HD 209458B *The fires of its nearby star are blowing its atmosphere into a tail.*

more than a couple hundred. The finds that have been confirmed include a totally unexpected solar system with at least six planets, each more massive than Earth, five of which are packed into orbits smaller than Mercury's.

It's an astonishing leap in the science of planet hunting, and it almost didn't happen: The Kepler mission was proposed to NASA no fewer than five times, beginning in 1992, by Bill Borucki, an astronomer at the NASA Ames Research Center in California. The first four times it was rejected on the grounds that it was simply too hard. Borucki's idea was to monitor a group of relatively nearby stars around the clock for several years. If any of them had planets, and if things were aligned just right, the orbiting worlds should regularly pass in front of the stars, in what astronomers call a "transit." The star's light would dim just a tiny bit, revealing not just the existence of the planet, but—since a big body blots out more light than a small one—how large the planet was.

NASA wasn't sold. But while Borucki may seem mild-mannered, he has the tenacity of a bulldog. "Bill has this personality where negativity just rolls off of him," says Natalie Batalha, the project's deputy principal investigator. "He gets rejected, and they tell him that this proposal is bad, and he doesn't take it the way most people would."

That's commendable, but it's hard to blame NASA for being nervous: Borucki wasn't just claiming Kepler would find planets (assuming they were out there). He was claiming it would be sensitive enough to find Earth-size planets. They're far tougher to detect than the Jupiter-size giants everyone else had been discovering—but far more important: "Mankind," says Borucki, "needs to find out if there's life out in the galaxy. One of the steps is finding planets, and the next step is finding Earths."

In the past 20 years scientists have found roughly

500

planets orbiting distant stars.

A few may have been spotted already. No fewer than 68 of Kepler's 1,200-plus candidates are approximately Earth-size, and while most of these possible Earths are too close to their stars and thus too hot, five lie in the so-called habitable zones—the orbit where temperatures may be suitable for life. The only catch: The stars in question are much dimmer and cooler than the sun. That doesn't rule out life, but it does add a complication.

It's too early for Kepler to have found a true twin of Earth orbiting a sun-like star, even in theory. That's because mission guidelines require at least three successive transits—and thus, three full orbits—before anyone can claim a discovery. A twin of Earth would take a year to orbit its star, by definition. But Kepler has been in orbit only 18 months, and it takes so long to process the data that the candidates announced in February come from just the first four months of observations.

Even when the three-orbit milestone is crossed, astronomers won't be fully convinced they've found a second Earth until they can be certain the new planet is made of similar materials to our own world. Planetary theorists had long assumed that if a planet was the size of Earth, it must, like Earth, be made of rock, some iron, and a bit of water. Maybe not, though: A lot depends on what was in the cosmic cloud that gave birth to star and planet alike. If the mix was much different from our own ancestral cloud, you might get an Earth made mostly of carbon, with a diamond core. Or it could be made largely of ice. It could even be a humongous droplet of water, 8,000 miles across.

Those bizarre possibilities can be ruled out (or in) by gauging the planet's mass, and thus its density. That tells you what it's made of. The fluffiest planet Kepler has found so far has the density of Styrofoam; another, found in a ground-based search, could indeed be mostly water.

Even if Kepler does find a second Earth, there's no guarantee it will harbor life. If it does, that life probably won't have radio transmitters to give it away. Chances are, the first aliens we find will be microbes, and across interstellar space, the only way to find them is to look for how they might have altered their planets' atmosphere.

Doing that requires capturing a direct image of a distant Earth. That's currently beyond our abilities—but thinking about future technology is something that astronomers do a lot of. Their dream: a space-based telescope called the Terrestrial Planet Finder, or TPF, that could take such a picture. The big hurdle is to blot out the star itself, which would overwhelm the feeble light of an adjacent planet—either by building a sort of mask into the telescope itself or, more fantastically, by positioning an "occulter" thousands of miles away in space and lining it up the way you might hold up your hand to block out a flashlight shining into your eyes.

All this will be excruciatingly difficult to pull off and not especially cheap. NASA originally wanted to launch TPF sometime during this decade, but budget pressures have put that off into the indefinite future.

So for the near future, anyone who's interested in alien life will probably have to be content with simply finding out how many Earths are out there. If they exist, Kepler is going to find them, and it's becoming clearer every day that it will find them in numbers far greater than anyone ever suspected. Kepler stares at only about 156,000 stars. It has now found candidate planets around roughly 100 of them—a whole lot more than the one in a million Jodie Foster talked about. Not only that, but Kepler's field of view spans about 1/400 of the entire sky, so there must be far more planets that it has no way of seeing. And because it's a modest-size telescope, it looks only at nearby stars.

It's hard, in short, to overstate the importance of this relatively modest space probe that might easily have never gotten off the ground. "Over the next couple of years," says David Charbonneau, a Harvard planet hunter who spends some of his time working with the Kepler team, "Kepler is going to revolutionize what we know about planets in the Milky Way."

SKY FIRE *Comet McNaught, photographed from Australia in 2007*

The Comets That Came From Far Away

Astronomers have been convinced for decades that the comets were born during the earliest days of the solar system, in the icy darkness around Neptune and beyond. Now, however, there's a new theory: Some of our solar system's comets were not in fact born here, research suggests. Rather, they were kidnapped from another star.

Most comets live in the Oort cloud, a vast haze of rock and ice that surrounds the solar system. Sometimes the Oort loses its grip on a comet, and it comes tumbling in toward the sun; those few escapees are the ones we see. Unfortunately, says Hal Levison, an astronomer at the Southwest Research Institute in Boulder, that theory doesn't quite add up. The Oort cloud is too dim to see directly. Instead, scientists use the rate at which comets plunge sunward to calculate the cloud's overall population, and the conventional story simply can't build a cloud the size of the one that's out there. It's just too big to have all been birthed by our proto-sun.

A new theory developed by Levison suggests that the solar system wasn't born alone, but instead as part of a litter of hundreds of stars, packed together inside a huge cloud of gas. Under those conditions comets would have been yanked and passed from one star to the next, often winding up far from their first home. At most, says Levison, we're likely to have seen only one locally born comet in the history of human observation. Which means that the next time a visible comet appears in the night sky, it's almost certainly a beautiful immigrant—a cosmic lesson perhaps worth heeding on Earth.

SOME OF OUR SOLAR SYSTEM'S COMETS WERE NOT IN FACT BORN HERE, THE RESEARCH SUGGESTS. RATHER, THEY **WERE KIDNAPPED** FROM A NEARBY STAR.

The Solar System's Quiet Spots

If you're looking to travel a long distance and yet go nowhere at all, you couldn't do much better than the Lagrange points. Our solar system may be a place of constant activity—of wheels within wheels as planets orbit the sun, moons orbit planets, and comets and asteroids weave through all the clutter—but out there somewhere are a few places where traffic comes to a grand standstill.

Those quiescent spots were discovered—or inferred—in 1772 by French mathematician Louis Lagrange, who realized that gravity does not pull with equal force everywhere in the cosmos. Traveling from the Earth to the moon, for example, a spacecraft is constantly tugged backward by terrestrial gravity until it completes about 80% of its journey, after which the much weaker gravity of the moon takes over. The ship essentially goes from climbing uphill to falling down. Just at that transition point, the pull of the two bodies essentially equalizes, and an object could simply hang in space.

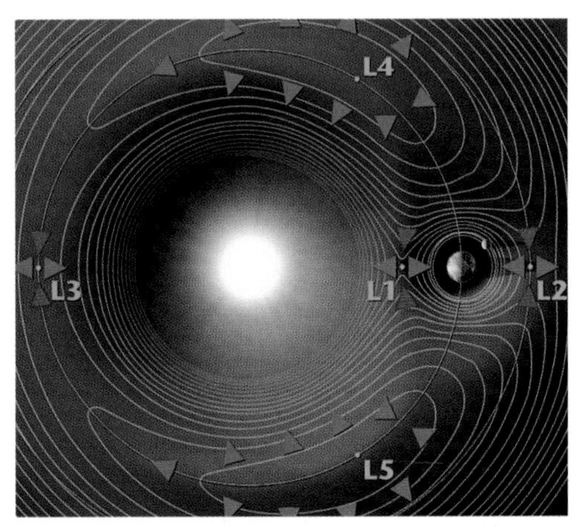

ALL IS QUIET *A map of Lagrange points one through five shows the spots in space where gravity achieves balance.*

There are numerous Lagrange points that have been mapped—between the Earth and the sun, trailing or leading Earth as it orbits the sun, and 180 degrees opposite Earth on the other side of the sun. In all these places a spacecraft would, technically, be moving, but since it would pace Earth precisely, it would appear to be stationary. That makes Lagrange points more than mere curiosities. As NASA seeks to redefine its ambitions in space, the Lagrange points are shaping up to be intriguing spots from which to conduct cosmic observations, unperturbed by earthly light, gravity, or radio interference. It's not the stuff of one small step and boots in the dirt, but it could be a great way to do smart science—which is what space travel is supposed to be about in the first place.

MOMENT OF IMPACT *This is how it might have looked as Deep Impact's 820-pound copper cannonball struck the flank of Comet Tempel 1 in 2005.*

A Second Dance With a Comet

Since 2007 scientists at NASA's Jet Propulsion Laboratory knew they wouldn't be doing anything terribly romantic on Valentine's Day 2011. That's because the Stardust spacecraft, which had already paid a call on Comet Wild 2 in 2006, would be making a second visit to Comet Tempel 1. On Valentine's night, precisely on schedule, Stardust flew past the comet at a distance of just 110 miles, bringing to a close a mission that was as much of a navigational triumph as it was a scientific one.

NASA had barnstormed Tempel 1 once before, back in 2005, when the Deep Impact spacecraft flew by and fired an 820-pound copper cannonball into the flank of the comet in order to blast out a plume of debris that the spacecraft could scan and analyze. Stardust, meantime, had had its own encounter with Comet Wild 2 in 2004, but after that was done, it was still fit for flight. NASA later decided to repurpose it for a flyby of Tempel 1. Not only might this provide a peek inside the Deep Impact crater, it could also allow scientists to study how the comet's surface changed over the course of a few years. All of this would provide important clues to comets' composition.

It took thousands of tweaks of Stardust's thrusters to make the rendezvous happen, but when it did, the results were worth it. The ship snapped 72 pictures in its brief flyby, and they showed that the comet's surface had indeed been resculpted significantly in just a very short time—at least on the cosmic clock. NASA will spend years studying the data further, and that too will be energy well spent. Comets are among the oldest artifacts of the ancient solar system; a good look at them is a good look at our very origins.

THE COMET
TEMPEL 1
IS NINE MILES LONG AND THREE MILES WIDE. IT TRAVELS AT 66,880 MPH, CIRCLING THE SUN BETWEEN THE ORBITS OF MARS AND JUPITER ONCE EVERY 5½ YEARS.

WORLDS WITHIN WORLDS *Saturn's oversize moon Titan is dwarfed in this image of the planet itself and its elegant system of rings.*

Is There Life on Titan?

The Cassini spacecraft has been orbiting Saturn since 2004, but a major focus of its mission has always been the planet's largest moon, Titan. Bigger than Mercury, Titan is also the only moon massive enough to have a significant atmosphere—one rich in organic compounds, including methane and other hydrocarbons. Titan's frigid temperature precludes life as we know it, but it has long been thought to offer hints about what a planet might look like before biology emerges.

Investigators have, however, pondered one other remote possibility. If an alternative form of life could exist with liquid methane taking the place of water in cells—something that's theoretically possible—Titan is just the sort of place it might thrive. About five years ago NASA astrobiologist Chris McKay suggested that if such "methanogens" existed, their collective metabolism could lead to lower levels of ethane and acetylene in Titan's atmosphere than you'd expect and a tendency of hydrogen to migrate down to the surface, where the organisms would be sucking it in as food.

Scientists already knew there wasn't as much ethane in Titan's atmosphere as they'd expected, but in 2010, Cassini found that the other two conditions had been met as well. So, life on Titan, right? Well, not so fast. There are other, easier explanations for the findings, including sampling errors or some other factor that's causing the hydrogen to migrate. But McKay won't rule out the life explanation. In the hedged language of the scientist, he says that Cassini's findings are "consistent with" life. That's not the same as "evidence for" life. But it's not evidence against either.

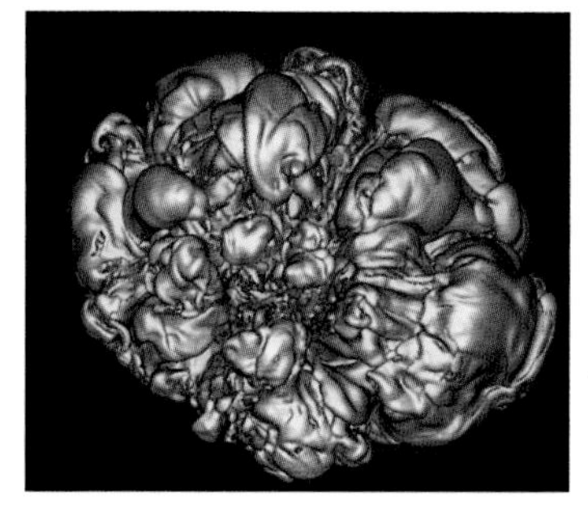

IGNITION *A computer simulation shows what would take place in the first half-second of a supernova explosion, as matter erupts from the core of the star.*

The Secrets of the Supernova

Most stars end their lives in a whimper, but the truly massive stars go out with a bang—a supernova. As familiar as most folks are with the idea of a supernova, astronomers still haven't completely figured out how they work. Thanks to a new computer simulation, physicists at Princeton and Lawrence Berkeley National Laboratory are getting closer.

When a star exhausts its nuclear energy, it collapses in on itself. Even such a massive cave-in is not enough to produce a shock wave of the supernova's magnitude. Instead, physicists believe an extra kick is provided by a blast of subatomic neutrinos that blow outward from the interior. The problem is, neutrinos are so ethereal they ought to pass through the star without perturbing it. The new simulation, however, reveals that the shock wave that does emanate from the star churns so violently that it gives neutrinos more of a chance to mingle with the stellar matter and transfer their considerable energy. The result: the signature flash of a dying star visible halfway across the universe.

CLOCKWISE FROM RIGHT: DAVID MCNEW/GETTY IMAGES; ADAM BURROWS/PRINCETON UNIVERSITY—HANK CHILDS-VACET/VISIT; NASA

What's on ET's Menu? How About Arsenic?

Not long ago, a team of scientists discovered what may be the oddest bacteria on Earth—and that has real implications for the possibility of life in space. The microbes, which caused a huge stir in the space community when the paper reporting them was published in the journal *Science*, live in hellish conditions in Mono Lake, a super-salty, alkaline, arsenic-rich body of water in eastern California that would be toxic to most organisms. But the new bug doesn't just thrive there: It uses arsenic in place of the standard phosphorus as a building block for its internal proteins and DNA—and nothing like that has ever been seen before.

Scientists who hope to discover alien life someday have always had to concentrate on life as we know it—because after all, if we don't know it, how can we know where and how to look? It's never been shown that other kinds of life are impossible, though. For decades, for example, biologists have wondered if silicon could take the place of carbon, the basis of all life on Earth, in forming self-reproducing, information-carrying molecules like DNA. Indeed, Arizona State University physicist Paul Davies, a co-author of the new paper, has argued that life as we don't know it might exist as a sort of "shadow biosphere," even here on Earth.

The new discovery proves that that isn't just wild speculation. If alternate biologies can happen here, there's no reason to think they can't happen elsewhere—in many more forms than just the new, arsenic-loving kind. If so, it's great news for astrobiologists: The more variety out there, the more places you can go looking for aliens. "We've cracked open the door for what's possible for life in the universe, and that's profound," says lead author Felisa Wolfe-Simon of the NASA Astrobiology Institute and the U.S. Geological Survey. "What else might we find?"

100

THAT'S HOW MANY MILLIONS OF SPECIES MAY EXIST ON EARTH. ALL WERE THOUGHT TO USE PHOSPHOROUS IN THEIR DNA. ONE IS NOW KNOWN TO USE ARSENIC INSTEAD.

NO PLACE TO LIVE *Not for normal bacteria, that is. But the briny waters of California's Lake Mono are perfect for organisms that live on arsenic.*

OPEN FOR BUSINESS *An artist's rendering of the Herschel Space Telescope as its Ariane 5 booster was carrying it into space in May 2009*

11.5 feet

THE DIAMETER OF HERSCHEL'S MIRROR, THE LARGEST EVER BUILT FOR USE IN SPACE

The Telescope for Invisible Stars

It's no secret that space is cold. But in some places it's so frigid that light can't radiate in the visible spectrum, which makes celestial bodies invisible. Now the Herschel Space Observatory is exposing them. Launched in May 2009 by the European Space Agency, Herschel scans the skies in the infrared spectrum. In order to avoid infrared interference and temperature fluctuations from Earth, it hovers in space at the second Lagrange point, about 930,000 miles away from Earth, but pacing the planet in its orbit around the sun. Herschel will operate for at least three years, during which time it will watch stars and planets being born, revealing more about how the universe came to be. Herschel is equipped with a mirror 11.5 feet in diameter, the largest ever built for use in space.

In the short time it's been in space, Herschel has done some extraordinary work. It has helped explain the origins of so-called starburst galaxies, the most prodigious starmaking factories in the cosmos. It has also taken the most detailed images ever of the relatively nearby Andromeda galaxy, studying in particular a massive ring of dust in Andromeda's center, measuring 75,000 light-years across. The curious formation is thought to be the result of an eons-ago collision with another galaxy.

Japan: The New Space Pioneer

The country that invented the Walkman may be reclaiming its image as a tech pioneer. In 2010 a revolutionary spacecraft built and launched by JAXA, the Japanese space program, successfully unfurled the world's first solar sail—a spacecraft that uses the thrust of charged particles streaming from the sun to propel it. Just three days later Japan announced that a ship that left Earth seven years ago had returned home, carrying with it a soccer-ball-size pod built to collect the first fragments of an asteroid ever brought to Earth. The solar sail may look low-tech, but its mission could help make future trips to the outer reaches of our solar system possible. The asteroid probe, meantime, achieved the first ever roundtrip to a celestial body other than the moon. With a moribund economy, the JAXA accomplishments have given Japan a reason to look away from its earthly problems and get satisfaction from the sky.

HOMECOMING *A pod carrying asteroid bits is recovered in Australia.*

FROM TOP: ESA; JAXA/AFP/GETTY IMAGES/NEWSCOM

MORE THAN IT SEEMS *The spiral galaxy Messier 83 is dazzling to look at. But what you can't see is the invisible blob of dark matter in which it sits.*

Dark Matter and Bright Galaxies

Ninety percent of the universe is missing. The AWOL material is called dark matter, an invisible substance that allows galaxies to spin as fast as they do without flying apart. Just about every galaxy, including our own, is held inside a blob of dark matter like a butterfly in a glass paperweight.

But not all galaxies are the same. Some are pipsqueaks, some are giants, and some are overachievers, churning out 1,000 new suns a year for their first 100 million years. These starburst galaxies have long puzzled astronomers, but new findings from the University of California at Irvine explain them. The answer, again, is dark matter.

If the dark-matter blob in which a galaxy formed is too big, the scientists now reason, hydrogen couldn't fall together efficiently enough to sustain a starmaking frenzy. If the blob is too small, the hydrogen falls together too efficiently, causing a frenzy—but a short one. It is midsize blobs that create starburst galaxies. This insight came courtesy of the Herschel Telescope, which detects infrared radiation. Young, far-off galaxies emit especially large amounts of infrared. The brightest spots in Herschel images represent the densest clots of galaxies. The scientists compared these pictures with computer simulations of the early universe, which reveal how dark matter should have been distributed. This produced a good match between medium-size lumps of dark matter and starburst galaxies. In other words, the ancient model is consistent with the current reality—meaning the universe's least visible matter yields it most brilliant visible galaxies.

IT JUST AIN'T SO ...

MOON JEWELS *Volcanic glass beads, like these collected by Apollo 15, have mixed with lunar permafrost.*

The Dry Moon: Wetter Than You Knew

The lunar surface is nothing if not predictable. You've got dust, you've got rocks, you've got bigger rocks. One thing you don't have is water—which has always made the idea of homesteading the place a challenge. As it turns out, however, the moon is a lot wetter than we ever knew. NASA's LCROSS (Lunar Crater Observation and Sensing Satellite) mission made that discovery when it crash-landed a spent rocket booster near the moon's south pole and directed the LCROSS satellite itself to analyze the debris plume that resulted.

It was no news that there was water vapor in the plume—the lunar poles are home to at least traces of permafrost, which is thought to have been carried in by crashing comets. Much of the ice would evaporate as soon as the sun struck it, but any that was buried or landed in permanently shadowed craters would last forever. The real surprise from LCROSS was just how much of this dusting there was—about 50% more than astronomers had anticipated. That makes the moon about twice as wet as the Sahara Desert. Okay, twice as wet as the Sahara is not exactly beachfront property, but it could be enough to allow future settlers to manufacture their own water supply on-site, which would be easier and cheaper than hauling it from Earth. Water not only could be used to sustain astronauts but could also be broken down into its constituent oxygen and hydrogen to be used as rocket fuel and breathable air.

The Mind

ADS THAT TARGET OUR EARS ■ MOM'S DEPRESSION AND BABY'S WEIGHT ■ MARIJUANA AND SCHIZOPHRENIA
MIND READING AND CRIME ■ GAYS, GENDER, AND FACIAL RECOGNITION ■ YOUR BRAIN'S BROKEN HEART
REWRITING YOUR PHOBIAS ■ MEET THE SUPERTASKERS ■ THE ANTI-AMNESIA CAMERA

Alzheimer's Unlocked

After years of disappointing vaccine and drug trials, researchers are finding new ways to interrupt the memory-robbing disease, just in time for an anticipated explosion in cases.

By Alice Park

Not all of Dr. Richard Mayeux's elderly patients have Alzheimer's disease; not all will even go on to develop it. Most of them are still leading full, healthy lives. But Mayeux, an Alzheimer's researcher and physician at Columbia University, asks them all anyway: Will they help him in his war against the disease? It's a war that badly needs winning.

More than 5 million Americans currently suffer from Alzheimer's disease, a number that will grow to 13.4 million by 2050. Health experts estimate that a 65-year-old has a 10% risk of developing Alzheimer's and that baby boomers currently approaching peak age for the disease (60 to 80) will add $627 billion in Alzheimer's-related health care costs to Medicare.

Mayeux knows that defeating Alzheimer's means first recruiting volunteers to join a study that can help identify who is at greatest risk of developing the condition. The results could paint a clearer picture of the factors that put people in danger. A disease that gives up clues to those factors is one that has revealed its weak spots. Still, the patients take some convincing. Said one prospective participant: "The way I see it, even if you predict when I will get Alzheimer's, you haven't got anything that I can do for it."

The patient has a point. Who in his right mind would want to know he had a disease that would inevitably rob him of that mind? Over time, the feeling has taken hold that beating Alzheimer's is the cold fusion of medical research: Everyone agrees it would be great, and everyone who tries it fails. And yet maybe, just maybe, that's changing. For the first time since the disease was identified more than a century ago, doctors are closer to uncovering its secrets. Alzheimer's, like all other degenerative ills, is driven by genes, and in the past year scientists have come up with a suite of relevant ones. The disease is

PHOTOGRAPHS BY GREGG SEGAL

THE SLOW FADE *Emily Bernice Chapman, 79, of Philadelphia is a 76ers fan who had an encyclopedic knowledge of her team. Dementia is robbing her of that—and more.*

thought to be caused by a buildup of protein-based plaques in the brain, and investigators now believe they have an understanding of possible ways to interrupt that process. Technology is helping too, as researchers exploit new ways to scan the brain and detect the first signs of trouble, pinpointing the very molecules that give rise to the disease.

The fact that there's renewed optimism concerning a disease that has long been the graveyard of hope comes mostly from scientists' ability to apply two important lessons learned from the disappointments of the past. The first involves timing. Experts are now convinced that it's crucial to treat Alzheimer's patients as early as possible, perhaps even before they show signs of cognitive decline, rather than attempt to improve a brain already scourged by the disease. The second involves the scope of the medical assault—adopting a multipronged approach that addresses as many of the disease's complex abnormalities as possible.

Shifting the focus to the earliest stages of the disease wasn't as obvious as it seems in hindsight. Cognitive decline is a natural consequence of aging. It's understandable, then, if doctors are reluctant to introduce more uncertainty by attempting to tease apart Alzheimer's dementia from the so-called senior moments typical of normal aging.

So rather than make the attempt, they focused on the most obvious target: the buildup of a protein called amyloid in the brains of Alzheimer's patients. While amyloid in living patients can be detected with a spinal tap, its presence doesn't necessarily indicate the disease; it's the accumulation of the protein into plaques, which also include cellular debris like dead and dying neurons, that is linked to the disease's symptoms. Initially it made sense for researchers and drugmakers to focus on finding ways to shrink plaque buildup; that, surely, would lead to improvement.

But to date, these well-intentioned efforts have been fraught with failure and riddled with side effects. The agents that target amyloid plaques affect other, healthy processes in the body too. What's more, in both human and animal trials it's been unclear whether reduction in plaques has any effect on brain function at all.

So perhaps amyloid isn't a critical contributor to the disease at all, but a red herring. Or perhaps amyloid is a factor in the pathology but only one of many. It's also possible that amyloid is indeed pushing the disease but that the vaccine and drugs used to dissolve the plaques were introduced too late and in too small a dose. If that's the case, then testing drugs on patients whose brains are just beginning to accumulate amyloid might yield more success.

In 2004 the National Institute on Aging (NIA), part of the National Institutes of Health, partnered with pharmaceutical companies to create the Alzheimer's Disease Neuroimaging Initiative, a $60 million project tasked with identifying easily detectable differences—preferably through blood tests or brain scans—between Alzheimer's patients and unaffected individuals. It was nuts-and-bolts science, unglamorous but essential, and it wound up attracting 600 patients who either already had symptoms of Alzheimer's dementia or had mild cognitive impairment, as well as 200 cognitively normal control-group volunteers.

The program has isolated a few dozen intriguing protein markers in blood and spinal fluid that may herald Alzheim-

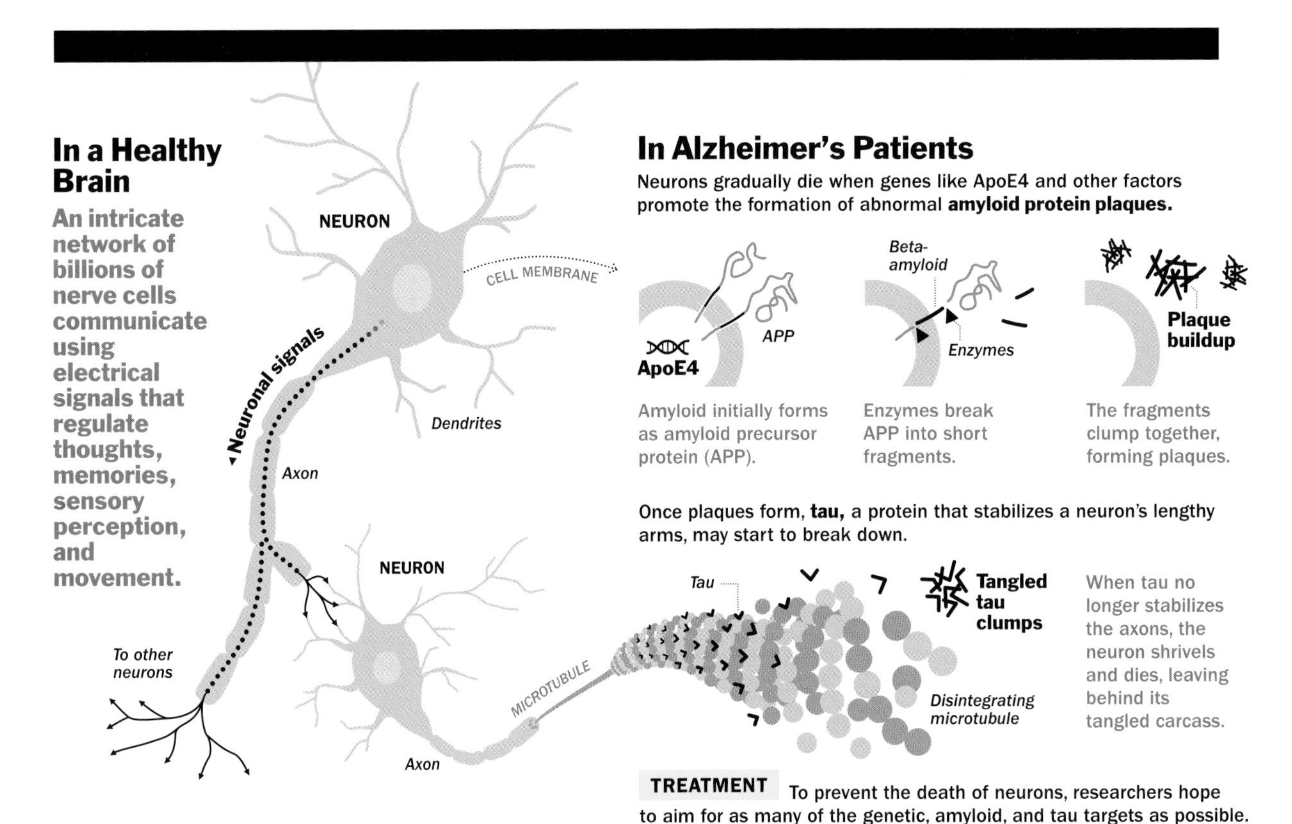

THE ARC OF TIME *Bill and Carole Bates have been married for 57 years. She now lives in an Alzheimer's facility, where he is a constant visitor.*

Scientists are peering deeper into our neural tissue to identify what drives the memory-robbing disease.

er's disease. Also, newer, better brain scans are helping detect the amyloid patterns that previously could be verified only by autopsy. Being able to say, "This patient appears to have Alzheimer's"—as opposed to "This deceased patient had Alzheimer's"—is no small thing. Still, as with the blood and spinal-fluid tests, the challenge remains to understand the link between plaques and actual symptoms.

For those answers, scientists need to test the measures on at-risk, asymptomatic populations. And for that, they need a consistent way to identify those populations. That's why the NIA and the Alzheimer's Association decided to update their criteria for helping doctors diagnose Alzheimer's by defining three distinct patient groups: those who are symptom-free but at high risk, those with mild cognitive impairment, and those with Alzheimer's dementia. The guidelines fold in the latest understanding of how brain scans and other tests can help distinguish among the three groups—helping physicians recommend better treatments for symptomatic patients as well as identifying the asymptomatic ones to serve as the healthy study subjects.

That's not to say the new diagnostic tools are perfect, and Mayeux, for one, is wary: "While everyone acknowledges that the [markers] are useful," he says, "there isn't yet a standard test that everybody agrees means the same thing when they see a score."

Still, the early data look promising, suggesting that the screens may be 80% to 90% accurate in picking up the earliest signs of the disease. This has more than simply diagnostic value; it also allows researchers to start targeting candidate medications and be more confident that the patients who receive them will benefit.

That, however, can't happen unless scientists start designing smarter therapies. It's clear that concentrating on amyloid alone is not sufficient to reverse Alzheimer's symptoms, so investigators are working hard to identify additional targets. Among the potential areas of interest are genes like apolipoprotein E (ApoE), which in certain forms can promote the formation of amyloid. Also attracting interest is a neural protein known as tau, which stabilizes axons, the long extensions that nerve cells send out to communicate with one another.

Even if therapies are years or decades away, identifying patients earlier in the disease cycle will remain valuable. It is a complete person who typically receives a diagnosis of Alzheimer's; it's the wreckage of that person that is eventually killed by the disease. The key is stepping in early enough to prevent the damage from being done.

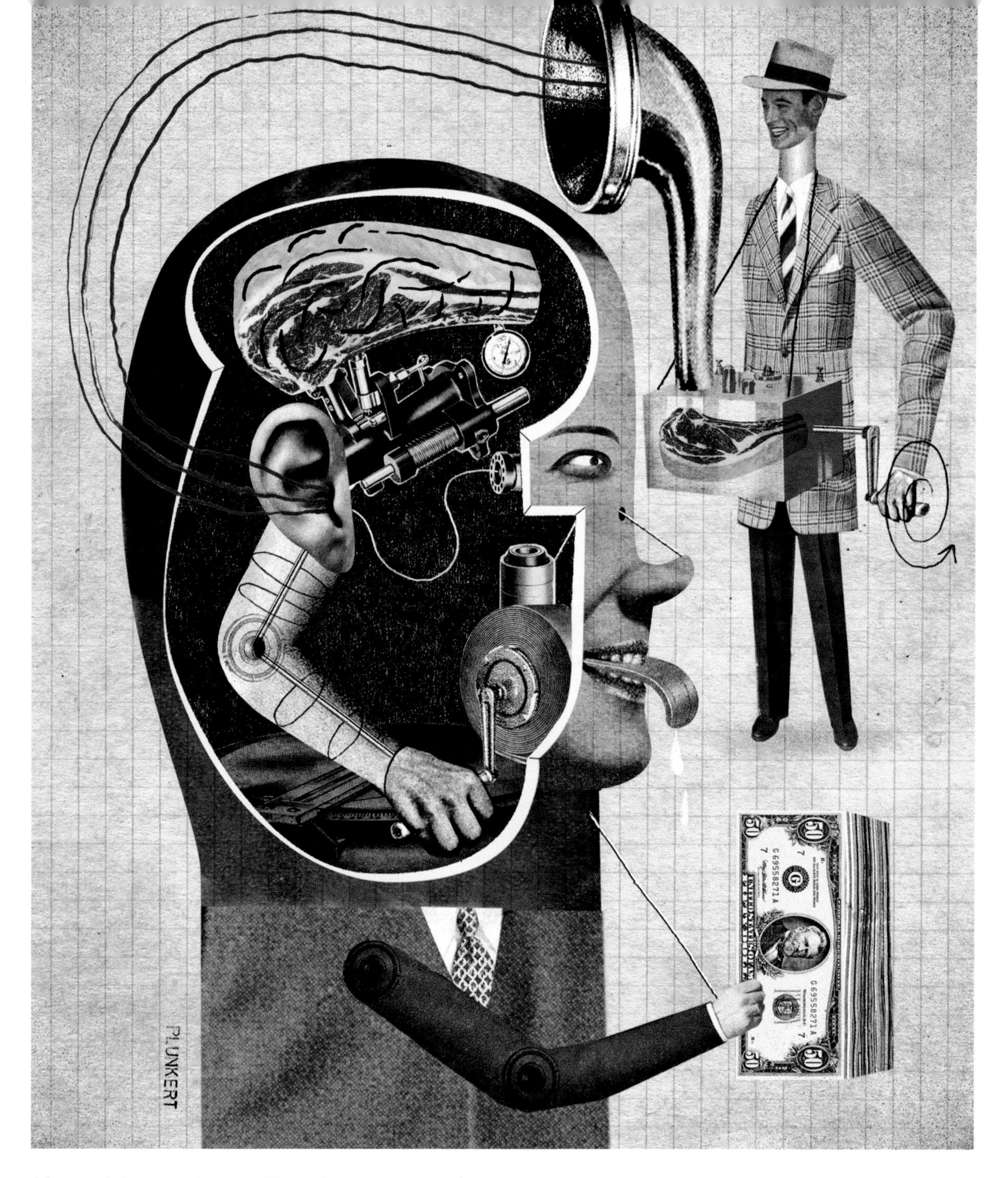

Neural Advertising: The Sounds We Can't Resist

We all tell ourselves we're way too smart for advertising—but that's hooey. Advertising works, which is why Madison Avenue is a $34-billion-a-year business. One of the most powerful ways ads reach us is through our ears, and Martin Lindstrom—author of the bestseller *Buyology*—is revealing just how.

Lindstrom has wired up volunteers so that he can monitor brain activity, pupil dilation, and more, and then played them recordings of familiar sounds, from commercial jingles to birds chirping and cigarettes being lit. The sound that blew the doors off all the rest—both in terms of interest and positive feelings—was a baby giggling. Other high-ranking sounds were the hum of a vibrating cellphone, an ATM dispensing cash, a steak sizzling on a grill, and a soda being popped and poured.

In all those cases the sounds already had meaning and thus triggered a cascade of reactions: hunger, thirst, happy anticipation. The advertisers just took advantage of that fact. TV commercials aren't the only places you'll hear such precisely targeted sounds. One Japanese department store has been designed as a series of soundscapes, playing sound effects such as lapping water in the sportswear section. European supermarkets similarly may begin piping in the sound of percolating coffee or fizzing soda in the beverage department or a baby cooing in the baby-food aisle. Oppressive as the auditory hard sell might be, there's not much you can do about it. When you're shopping, you can't simply change the channel.

FROM LEFT: DAVID PLUNKERT; LENNART NILSSON/SCANPIX; HAPA/GETTY IMAGES

Depression During Pregnancy Can Mean a Smaller, Less Healthy Baby Later

Women who suffer from depression or anxiety may be more likely to have underweight babies—even when those babies are born at full term, adding to a body of research that gives conflicting evidence about mental health and pregnancy. In a study that assessed 583 women in rural Bangladesh, 18% showed depressive symptoms, and 26% showed symptoms of general anxiety. Full-term babies born to the women with those conditions were about twice as likely to be born at or below a weight of 2,500 grams (or 5.5 pounds). The finding matters because small newborns are at much higher risk of health problems and infant death than babies born just a little bit bigger.

It's not clear why some studies find such a strong association between depression and anxiety and fetal development, while others do not. Perhaps mental-health problems in some populations are merely signals for other health problems that cause poor fetal development. If depression and anxiety are indeed the proximate causes of the babies' low weight, a number of mechanisms could be at work. Depressed women may be less likely to seek out adequate prenatal care while pregnant, they may fail to gain sufficient weight themselves, or perhaps the neuroendocrine effects of depression and anxiety throw off the fetal development process. Whatever the reason, if you're depressed—during pregnancy or not—seek help.

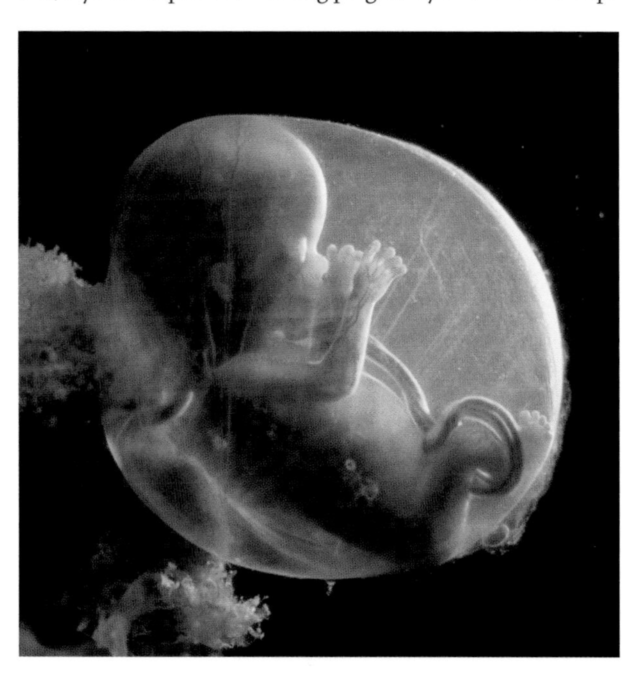

FOR DEPRESSED PREGNANT WOMEN THERE IS A

12%	15%
GREATER INCIDENCE OF PREMATURE BIRTH.	GREATER INCIDENCE OF LOW BIRTH WEIGHT.

What Marijuana Can Teach Us About Schizophrenia

Since the days of *Reefer Madness*, scientists have sought to understand the connection between marijuana and psychosis. Data suggest that those who smoke cannabis are twice as likely to develop schizophrenia as nonsmokers. One widely publicized 2007 review of the research even concluded that trying marijuana just once was associated with a 40% increase in risk of schizophrenia and other psychotic disorders.

But here's the conundrum: While marijuana went from being a secret shared by a community of hepcats and beatniks in the 1940s and '50s to a rite of passage for some 70% of youth by the turn of the century, rates of schizophrenia in the U.S. have remained flat, affecting about 1% of the population.

New research has explored the issue in depth. One study found that schizophrenics who smoked pot had an earlier onset of the disease—about two years—than those who did not. Gender could account for a lot of that, since males generally have an earlier onset of schizophrenia and are four times likelier to smoke pot as teens than females.

Genetics could play a role too. A French study found that schizophrenics who were pot-sensitive had three times the number of close relatives with psychotic disorders than those whose disease was not affected by the drug. The answer may lie in the brain's endocannabinoid receptors, which react to cannabis and affect the dopamine system. Discouraging pot use in schizophrenics or people at risk is difficult; the drug is extremely popular with this group. But difficult does not mean impossible, and when sanity itself is on the line, it's worth a try.

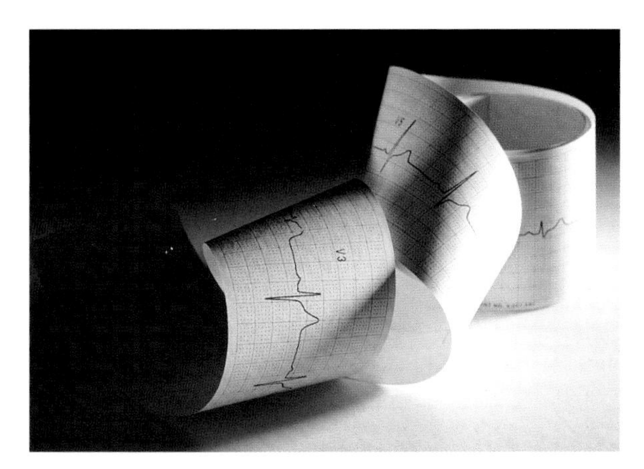

Fighting Crime by Reading Minds

What if we could read the mind of a terrorist? Researchers at Northwestern University in Chicago say they have taken a step closer to that reality with a test that could uncover nefarious plans by measuring brain waves.

In a study published in the journal *Psychophysiology*, psychologists used electrodes to measure brain waves of 29 undergraduates who had been told to mock-plan either a terrorist bombing in Houston in July or a vacation in a different city in a different month. The researchers then presented the students with the names of various cities, methods of attack, and dates. As they did so, they scanned their brains with electroencephalography, which records electrical activity. They watched for a particular brainwave—dubbed the P300 because it fires every 300 milliseconds—which signals recognition of something familiar. The P300's amplitude was large when the researchers presented the word "Houston," the city where the attack was planned.

The P300's potential as a method for confirming concealed information is limited. For instance, it becomes difficult to use if investigators do not already know the information they are trying to confirm. (How could they be sure of including Houston if they didn't know it was the target city?) Still, investigators believe P300 readings could be more reliable than lie detectors—an admittedly low bar—which measure such parasympathetic responses as respiration and sweating. Those can certainly be triggered by a lie, but also by any high-stress situation—including the mere experience of being interrogated.

THE P300

IS A SIMPLE BRAIN WAVE THAT TELLS A BIG STORY, SPIKING WHEN YOU RECOGNIZE SOMETHING—AND OFTEN REVEALING A LIE.

Facial Recognition: What It Reveals About Gender and the Brain

It's long been an accepted truth among married couples that it's usually the wife who must steer the pair through social gatherings, since she's the one who's far likelier to remember faces. In lab settings, women routinely outperform men in facial-recognition skills, both in terms of speed and reliability. Imaging scans have shown that part of the reason is that men rely only on the right side of the brain in summoning up images of a face they've seen before, while women recruit from both hemispheres—literally doubling their brainpower.

Now research from York University in Toronto has found that it's not just women whose brains are so nimble; it's gay men too. In the study psychologists recruited a sample group of homosexual men, heterosexual men, and heterosexual women. All the volunteers were shown pictures of 10 faces and given time to try to memorize them. Those 10 faces were mixed with images of 50 other people and flashed on a screen for milliseconds each. The subjects' job was to press a key when they saw a face they'd seen before.

When the results were tallied, the gay men and straight women scored equally well, and both did better than the straight men. Gay men, the researchers say, probably do so well at recognizing faces because, like women, they're putting both hemispheres to work at once.

The explanation is rooted in the genes. All people are born with coding that regulates body symmetry—including handedness, the whorl in the hair at the crown of the head and crosstalk between the brain's hemispheres. Gender and orientation can sometimes get mixed up with this, since a closely linked suite of genes controls them as well.

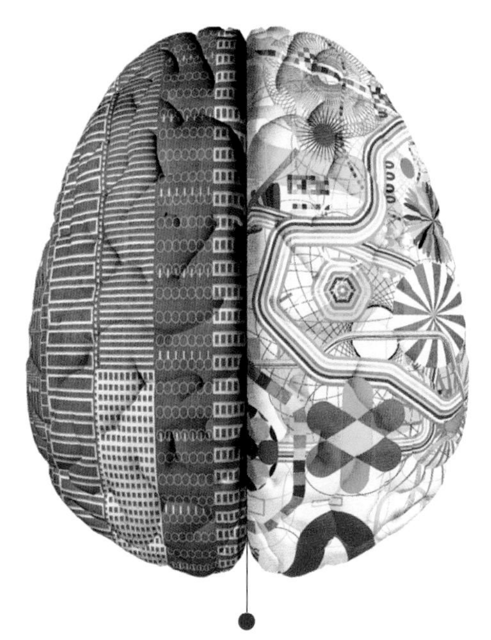

MEN RELY ONLY ON THE RIGHT SIDE OF THE BRAIN TO RECOGNIZE FACES, WHILE WOMEN RECRUIT FROM BOTH HEMISPHERES—LITERALLY DOUBLING THEIR **BRAINPOWER.**

FROM TOP: SHARIE KENNEDY/CORBIS; JOHN HERSEY

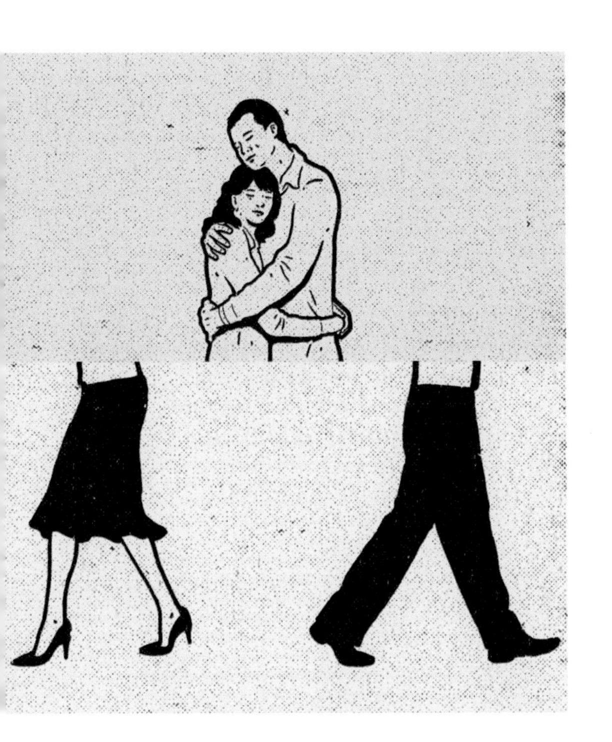

Phobias: How to Rewrite the Fear

If you're afraid of snakes or elevators or dogs or ponies or even the sound of another person chewing gum, you're not as alone as you think. According to the National Institute of Mental Health, 8.7% of Americans over the age of 18 have a specific—often rather peculiar—phobia. The standard treatment is called exposure and response prevention (ERP), and as the name implies, it involves deliberately, systematically exposing yourself to the thing you fear and then not responding by running away or avoiding or whatever else you typically do to extinguish the fear. ERP works, but it's not a lot of fun.

Researchers at New York University are trying a new approach. According to a study published in the journal *Nature*, when a person's phobia gets activated, there's a period immediately afterward when the traumatic memory that the phobia is based on becomes vulnerable. During that time, which lasts about six hours, you can reshape the memory—rewrite it in a way that removes the fear. The participants in the *Nature* study were first trained to fear a certain arbitrary stimulus—they were shown colored cards while receiving mild electric shocks—then reconditioned during the reconsolidation period not to make that association. The fear went away. It was still gone when the participants were retested a year later. For now, that's still in the realm of research, and ERP remains the best and most effective bet. But the more options there are for treating phobias, the more that frightened people will emerge and get well.

What a Broken Heart Looks Like Inside Your Brain

Say you're a college student who was just dumped by the person you thought was the one. You're moping around campus in your I've-given-up sweatpants and cursing the lover who made you feel so lousy. According to a new study, however, the object of your curses should really be your own brain.

The psychologists who conducted the work recruited 15 subjects still pining for their exes, hooked them up to a functional magnetic resonance imaging (fMRI) scanner and asked them to look at a picture of the person who had broken their heart—which the subjects themselves provided. After looking at the photo, they were asked to count backward from 8,211 by sevens, then look at another picture of a person they knew but were not in love with, then count backward again.

The fMRIs showed that the areas of the brain activated by the picture of the ex were the same ones associated with reward, motivation, physical pain, craving, and addiction. The photos of the strangers did not trigger those regions. Counting backward by sevens did help—temporarily. Permanent improvement requires time, but during that period it can be helpful to know that what feels like such operatic pain is also nothing more than simple neurochemistry.

8.7%

OF PEOPLE IN THE U.S. OVER THE AGE OF 18 HAVE A SPECIFIC—SOMETIMES RATHER PECULIAR—PHOBIA.

THERE MAY BE A SET OF BIOLOGICAL, GENETIC, AND PERHAPS BEHAVIORAL FACTORS THAT CONTRIBUTE TO **EFFICIENT MULTITASKING.** KIDS MAY BE BETTER AT DOING TWO THINGS AT ONCE THAN ADULTS ARE, SINCE THEY BEGIN TEXTING, WEB-SURFING, AND E-MAILING VIRTUALLY FROM THE CRADLE.

Never Mind Multitaskers; Meet the Supertaskers

Multitasking has become a way of life. Most of us think nothing of juggling a couple of chores at once, whether at home or in the office. But multitaskers are so last year; according to emerging research, there's a whole new—and very small—category of people who are nothing short of supertaskers. Most of us don't know a supertasker, but there are a few out there. During his presidency, Bill Clinton was famous for being able to work a crossword puzzle, play a game of hearts, read a briefing book, and talk to reporters all more or less at once. Scientists now suspect there may be a set of biological, genetic, and perhaps behavioral factors that contribute to efficient multitasking.

Daphne Bavelier, a professor in the Department of Brain and Cognitive Sciences at the University of Rochester, studies the effect of action-videogame playing on our ability to split attention and multitask. Bavelier has found that people who devote five hours or more a week to such action games for a year show heightened performance abilities that can verge on supertasking.

A year of videogames is not recommended for anyone, but the work does show something about the nimbleness of the brain, particularly in young children, who, thanks to advances in technology, seem to be naturally more adept at multitasking than previous generations. The researchers are especially eager to find out whether supertasking can be trained or learned in older folks as well. One place you should never multitask, no matter how attentive you think you are, is behind the wheel: In one study people talking on a cellphone took a full 20% longer to hit the brakes in an emergency than people not on the phone—more than enough to turn a near-miss into a fatal crackup.

FROM LEFT: CJ BURTON/CORBIS; MICROSOFT RESEARCH (12); TOM LANG/CORBIS

A Camera That Battles Memory Loss

To people with amnesia, the world is a mysterious place. For some who suffer from a type of memory loss that makes it impossible to recognize faces, even the reflection in the mirror requires an introduction.

Amnesiacs often retain a functioning procedural memory; if they knew how to play the piano before the injury or illness that cost them their recall, they'll still be able to. Likewise, they may retain a semantic memory, holding on to previously learned facts—Paris is the capital of France, for instance. What's lacking is an episodic memory—the ability to retain the sights and sounds of new experiences from one day to the next.

This is where a new memory-enhancing camera can help. Called the Sensecam, it hangs around a patient's neck and automatically takes photos with a wide-angle lens every 30 seconds. The patient can download the pictures later and review them in sequence. But why should a two-dimensional image of, say, a museum jog a memory of a recent visit? Neuropsychologists at Microsoft, the company that developed the camera, believe that its impromptu wide-angle images, which capture everything in the amnesiac's field of vision, provide a powerful memory cue—much stronger than a staged, traditionally framed picture of the same sight.

Microsoft is marketing Sensecam as a medical device, but there may be a bigger market than just amnesiacs. Since Alzheimer's disease targets memory structures first, cameras may provide a prophylactic, keeping memory intact longer—if not forever.

DIARY OF A DAY *Sensecam takes a new picture every 30 seconds.*

IT JUST AIN'T SO ...

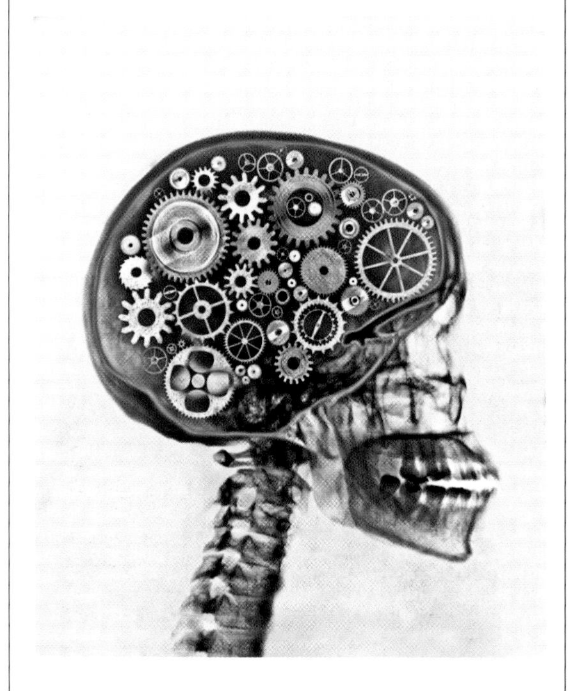

Fixing a Busted Personality

Psychologists dread patients with personality disorders. The narcissist, the histrionic, the antisocial are often more than simply troubled. They can also be absolutely convinced that there's nothing at all wrong with them and that the problem in fact lies with their friends, families, and doctors. A patient who denies the illness is awfully hard to fix.

That is particularly so in the case of borderline personality disorder (BPD)—a condition defined by volatile relationships, extreme sensitivity, poor impulse control, and a high risk of suicide or other forms of self-harm. Psychologists describe BPD patients as the psychological equivalent of third-degree-burn victims; they simply have no emotional skin.

Now, however, there's hope for BPD patients, thanks to a treatment called dialectical behavioral therapy (DBT). Dialectical behavior therapy is so named because at its heart lies the requirement that patients and therapists find synthesis in various contradictions, or dialectics. Therapists must accept patients just as they are (angry, confrontational, hurting) within the context of trying to teach them how to change. Patients must end the borderline propensity for black-and-white thinking, while realizing that some behaviors are right and some are simply wrong.

That can make for grueling therapy, but it works: a mix of tough love on the part of the doctor and learned wisdom on the part of the patient. Some 10,000 therapists are now trained in DBT. That's a huge corps of healers for the estimated 18 million Americans diagnosed with BPD—all of whom were once considered incurable.

Technology

ROBOTS AS ENGLISH TEACHERS ■ OVERCOMING PARALYSIS WITH GLASSES ■ AN INCUBATOR FOR THE THIRD WORLD
THE HYBRID CAR-PLANE ■ SPRAY-ON FABRIC DRESSES AND BANDAGES ■ STANDING ON YOUR OWN eLEGS
CHARGING WHILE DRIVING ■ A REAL IRON MAN ■ THE SELF-DRIVING CAR ■ WORMING INTO IRAN'S COMPUTERS

Battling Mosquitoes With Lasers and Genes

Malaria is a global scourge, and it's the mosquito that's the delivery system. Now two labs are taking aim at the pest in two creative ways—and neither involves bug spray.

By Jeffrey Kluger

It's been a bad year to be a mosquito. The world's most despised insect is responsible for 250 million cases of malaria a year and 1 million deaths, taking the life of a child every 43 seconds. That's not entirely the mosquito's fault. It didn't ask to subsist on an all-blood diet. It doesn't ask to pick up the malaria-causing *Plasmodium* parasite every time it bites an infected animal or person, passing the disease on to its subsequent targets. All it wants to do is fly, breed, and eat in the brief two-week lifespan nature gives it. Intentions notwithstanding, the mosquito does a lot of damage in that brief fortnight—and for that it must die, or at least change its ways. Mosquito eradication or control has been a goal throughout human history, but modern science is making it happen, thanks to two independent projects progressing along two parallel lines. One is using imaginative genetic engineering to let mosquitoes live but render them unable to spread malaria. The other will simply zap them out of the skies with lasers.

The genetic part of the anti-skeeter push is the product of a collaborative effort by the University of Arizona and the University of California at Davis. For all the persistence of the mosquito, the investigators knew that the insect had weak spots they could exploit. For one, any mosquito-control strategy would need to target only females, since they're the ones that feast on blood, which is essential for developing eggs. For another, though there are many species of malaria-causing *Plasmodium*, only two cause the most severe, relapsing cases. Focus your efforts on mama mosquitoes carrying one of those two particularly deadly species and you can get the most bang for your work. But once you've figured out which bugs to go after, you still

DAVID SCHARF/SCIENCE FACTION/CORBIS

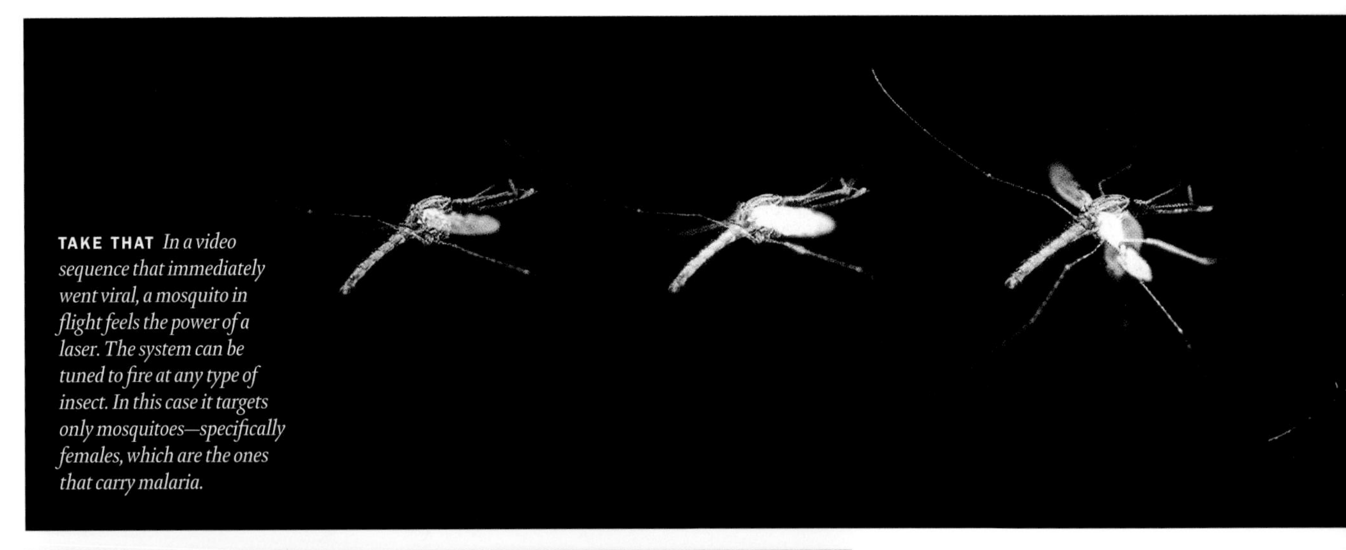

TAKE THAT *In a video sequence that immediately went viral, a mosquito in flight feels the power of a laser. The system can be tuned to fire at any type of insect. In this case it targets only mosquitoes—specifically females, which are the ones that carry malaria.*

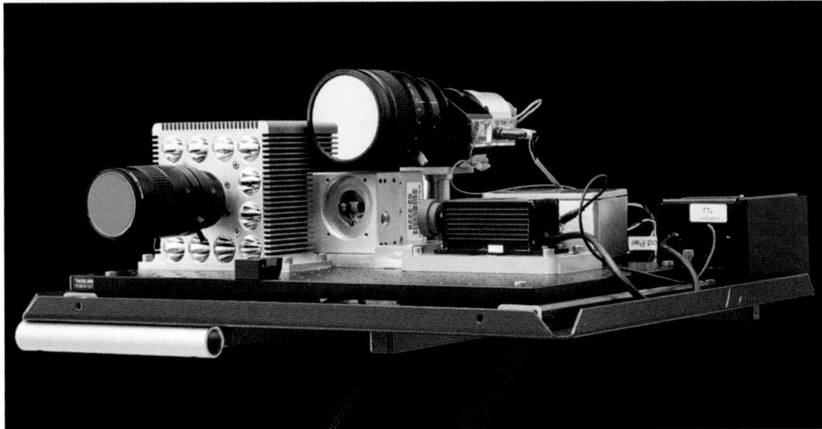

DEADLY BOUNDARY *The prototype mosquito laser (left) could give way to a self-contained fence-mounted system (right). A range finder (red light) would reflect off a target and bounce back (purple) with its coordinates. The laser (green) would then fire. The system could create a fence of light around homes, villages, and hospitals (far right).*

must pick the most vulnerable part of its genome. Here, too, there is a promising avenue.

A key player in the mosquito's metabolism is an enzyme known as Akt, which has a role in a host of functions, including immunity, antioxidant production, cell damage repair, growth rate, and lifespan. Manipulate the first three of those and you could help the mosquito neutralize the *Plasmodium* parasite. Tinker with the last two and you might shorten the insect's life—a very handy thing.

"In the wild, only the oldest mosquitoes are able to transmit the parasite," said Michael Riehle, an associate professor of entomology at the University of Arizona and the leader of the study. "If we can reduce the lifespan of the mosquitoes, we can reduce the number of infections."

To that end, Riehle and his colleagues inserted a modified Akt gene into the eggs of lab-bred mosquitoes. The eggs were then shipped to the molecular biology lab at Davis, where they hatched into mature insects with the altered gene woven into their DNA.

The next goal was to feed the mosquitoes blood infected with plasmodium and see what happened—and what happened was extraordinary. When molecular biologist Shirley Luckhart analyzed the innards of the mosquitoes, she found that none of the parasite was able to survive in the midgut—not a trace. What's more, the mosquitoes—which were allowed to live out their entire lifespans before being opened up—did not make it all the way to their usual two weeks.

Of course, mosquitoes in California and Arizona laboratories are not the same as mosquitoes in the wild, where the bugs are still having their way with humans. The next step is to breed a whole generation of insects with altered Akt and release them into the field. After that, the malaria-resistant mosquitoes must somehow displace the infectious population, which would otherwise be free to continue its deadly work. For that to happen, the scientists have to engineer some kind of advantage into their lab-grown bugs—make them bigger, stronger, and more aggressive toward their own kind, anything that will allow them to outcompete the species already entrenched.

That will take a while, but the investigators are extremely optimistic, if only because the first part of their work has gone so well. "If you want to stop the spreading of malaria, you need mosquitoes that are no less than 100% resistant to it," says Riehle. "If a single parasite slips through and infects a human, the whole approach will be doomed to fail. In [our] group of mosquitoes, not a single plasmodium managed to form."

Less complex—and a lot cooler—than the Atk research

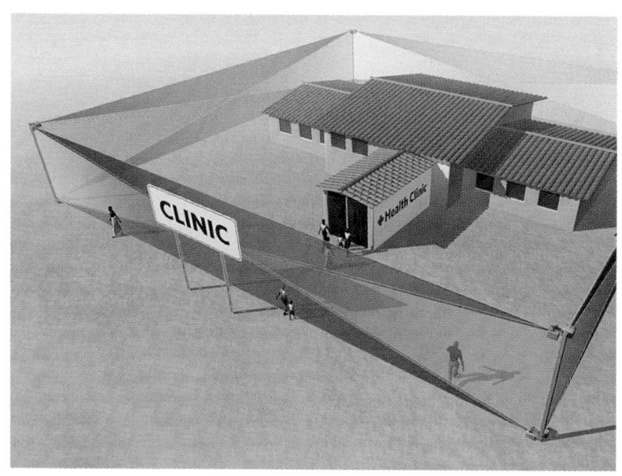

is the effort to develop an antimosquito laser. That improbable project is being spearheaded by Nathan Myhrvold, onetime chief technology officer for Microsoft and now head of the tech development company Intellectual Ventures. After years of working side by side with Microsoft chief Bill Gates, Myhrvold picked up some of Gates' own passion for global health and the issues that confront the developing world—particularly the malaria scourge—and decided to do something about it. At a 2008 brainstorming session with his Intellectual Ventures development team, the idea of an antimosquito laser system—a so-called photonic fence that could guard a home like a machine gun nest—came up. While it was perhaps the most outlandish suggestion raised that day, it was also one that turned out to be surprisingly achievable.

Lasers, after all, have been around for decades. And tracking systems that can point and aim them exist as well. The first major obstacle to an antimosquito system, then, would be cost—something Myhrvold and his team proved did not have to be an obstacle when they bought all their parts on eBay.

More important, the system had to be safe. It does nobody any good at all if a mosquito laser is overly sensitive, shooting at anything that moves—including birds, butterflies, and the odd passing child. The Intellectual Ventures engineers solved that problem too, tuning their laser system to react only to something that is the size and moves at the speed of a mosquito. In fact, they refined the system even further than that, specifically programming it to notice the signature wingbeats of females. Any other critter would pass by unmolested.

At a California technology conference in 2010, Myhrvold unveiled his invention, releasing hundreds of mosquitoes into a glass tank and letting his laser blast away at them. The audience roared, and the slow-motion video of the insects being shot from the sky instantly went viral.

Obstacles remain, of course. The system must still prove it can work in a less confined and controlled setting than a glass tank. Distribution, maintenance, and repair could all be issues too. The predicted price is coming down—to about $50 a unit if volume is high enough. One thing that won't be a problem is parts. Myhrvold is hoping to switch from a red or green laser to a blue one, both because blue lasers are somewhat safer and because Blu-ray hardware is now being built in cheap and affordable quantities for home DVD systems. Technology is getting nimble indeed when a device that does nothing but play movies in one part of the world can save lives in another.

Mr. Robot's English Class

Call it the job terminator. South Korea, which employs some 30,000 foreigners to teach English, has plans for a new addition to its language classrooms: the English-speaking robot. Students in a few schools started learning English from the robo-teachers late last year; by the end of this year the government hopes to have them in 18 more schools. The brightly colored, squat androids are part of an effort to keep South Korean students competitive in English—and if kids learn better when they like their teacher, the robot should be a winner.

Not all the English-speaking robots look the same. One, dubbed Engkey, is one meter (3.3 feet) tall and shaped like an egg, with a screen on top that displays a human face. Less adorable than other models, it is nonetheless hugely popular. "It is awesome and interesting," said one third-grader—in what is already nicely colloquial English—when Engkey was given a test run at Hakjung Elementary School. Despite the teacher's nuts-and-bolts anatomy, a real human was behind the instruction: The face shown on the robot's screen was that of a live teacher in the Philippines beaming in her lesson. That's decidedly lower tech than smart robots fluent in English and able to respond and correct others in real time, but those are in development too. Meantime, the kids—like all kids—are lovers of gadgetry and figure that any robot in the classroom is better than none. "I felt I could learn English better [with the robot]," said the third-grader.

CLOCKWISE FROM TOP LEFT: KOO SUNG SOO; JAMIE CHUNG; DESIGN THAT MATTERS; JAMIE CHUNG

Flick an Eye and Control Your World

How do you communicate when your brain is active but your body isn't? The EyeWriter—a collaboration from the Ebeling Group, the Not Impossible Foundation, and Graffiti Research Lab—uses low-cost eye-tracking glasses and open-source software to allow people suffering from any kind of neuromuscular syndrome to write and draw by tracking their eye movement and translating it to lines on a screen. The eye-tracking software focuses specifically on the position of the pupil—a precise target that makes for good control. Pupil-centric technology is not unknown; what is new in this case is the precision with which the EyeWriter works. LEDs that shine in the near-infrared are built into the glasses and illuminate the eye. That darkens the pupil and makes it easier for the tracking system to pinpoint its location.

Another asset that distinguishes the EyeWriter is its flexibility. The open-source software means that the same basic technology could be used for a wide range of purposes—drawing, writing, reading, controlling devices, and Internet surfing. The current model was created for Tony "Tempt1" Quan, an L.A.-based graffiti artist who was diagnosed with Lou Gehrig's disease in 2003. After trying the EyeWriter—the first time he'd drawn anything since he became fully paralyzed—Quan said, "It feels like taking a breath after being held underwater for five minutes."

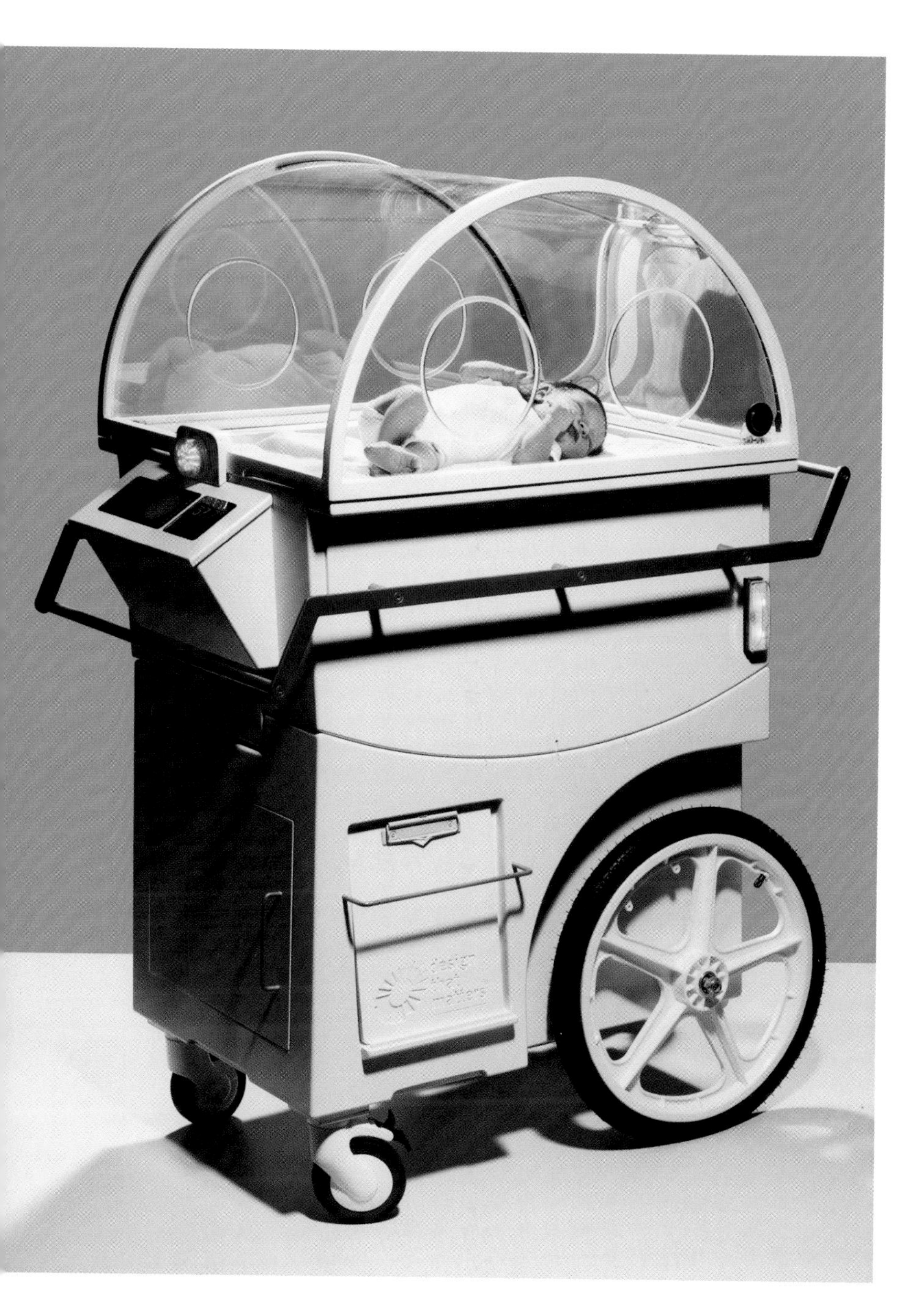

How Ingenuity and Spare Parts Are Saving Babies

The numbers alone explain the problem: Every year 4 million babies worldwide never live to their 1-month birthday; 1 million of those die the same day they're born. Child-health specialists believe about 1.8 million of the total could be saved if they could be kept sufficiently warm. The best way to ensure that, of course, is for mom to hold the newborn close—the kind of constant nurturing that's known as kangaroo care. But with more than half a million mothers dying each year from childbirth-related problems and many more unable to provide consistent care, that's not possible. Enter the NeoNurture incubator.

In the West, hospital incubators are as common as heart monitors—which is to say they're everywhere. But they also carry a hefty price tag—about $30,000 each, way too much for the developing world. The NeoNurture incubator, developed by university students in the U.S., gets around that problem, using an underutilized resource, old car parts, to build reliable incubators on the cheap. Headlights provide heat; a dashboard fan circulates air; a door-chime and signal-light assembly is rejiggered into an alarm system that alerts caregivers when things go awry with the heating mechanism.

The incubator is a two-component system: a chassis below and a detachable bassinet on top. The bassinet contains most of the life-support equipment (below left), allowing the baby to be moved about without the heavy cart and still remain warm and safe. The base is designed with rough terrain in mind; it has a wraparound handle that allows two people to lift and move the entire assembly over rugged ground—vital in places where the luxury of a modern hospital ward is unknown. The chassis also includes a storage cabinet for sterile blankets. The NeoNurture may well continue to improve, and why shouldn't it? The average SUV has about 40,000 parts, all of which could be repurposed for new—lifesaving—uses.

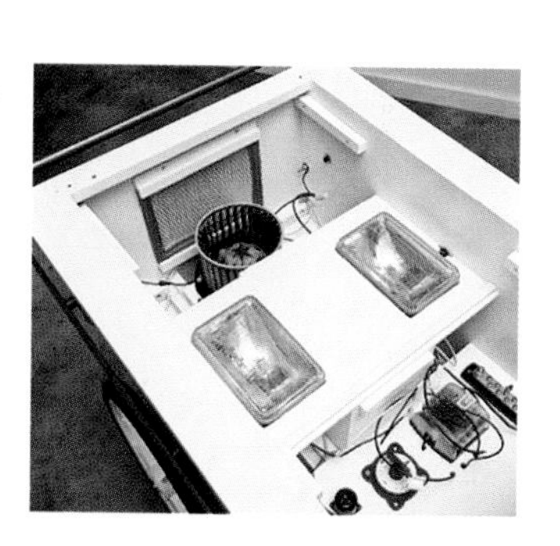

4 million

HOW MANY INFANTS IN THE DEVELOPING WORLD DIE EVERY YEAR WITHIN A MONTH OF BIRTH. HALF OF THESE NEWBORNS WOULD SURVIVE IF GIVEN A WARM AND CLEAN ENVIRONMENT IN WHICH TO GROW STRONGER.

FROM CAR ...

In car mode the 100-horsepower Transition gets 35 mpg.

The company wisely suggests that pilots open the gull wings only at an airport.

... TO PLANE

The plane needs a mere 1,700 feet of runway to achieve liftoff.

It's a Car, It's a Plane; No Need to Choose

The Terrafugia Transition could redefine the convertible—and the airplane and the automobile and the entire idea of how you get from place to place. Designed by a team of MIT aeronautics engineers, including Terrafugia co-founders Carl Dietrich and his wife, Anna Mracek Dietrich, the Transition is a street-legal, airworthy, airbag- and parachute-equipped flying car that, at $200,000, is priced lower than a Lamborghini. The first commercial models are already being built at a 19,000-square-foot production facility in Woburn, Mass. With its wings retracted like football goalposts, the Transition, whose 100-horsepower engine gets it 35 miles per gallon on terra firma, isn't going to be a match for an Italian sports car. But extend the vehicle's gull wings—and you are requested to do so only at an airport—and the rear-propeller-powered Transition can fly two passengers about 500 miles at a cruising speed of 105 mph. You will not be heading to the rental counter.

The plane is not only versatile but easily parkable and storable. It stands just 6 feet 6 inches tall, and in its car configuration it's just 90 inches wide. Its wingspan is 26.5 feet, but that won't matter since by then you're off the street and in the sky. The vehicle is a featherweight 970 pounds unfueled and 1,430 pounds loaded for takeoff. For all their imaginativeness, the Terrafugia designers also have something any start-up would love: true believers in high places. The U.S. military is interested in developing Terrafugia-type vehicles for the battlefield and has signed the company up to participate in the initial R&D.

TERRAFUGIA (6)

DON'T STOP WITH CLOTHES.
SPRAY-ON TECHNOLOGY COULD BE USED IN HOSPITALS AS BANDAGES, QUICK-CHANGE BEDDING, AND SLOW-RELEASE MEDICINE PATCHES.

Hissss ... a New Dress

Nothing to wear? No problem. Just crack open a can of couture and spray yourself a new dress. If you ever despair for the state of the world, consider that you live in an era in which clothing can now join cheese, insulation, and hair on the list of things that can be applied by aerosol.

That surprising development comes courtesy of the British company Fabrican, which has developed a way to bond and liquefy fibers so that textiles can be sprayed out of a can or spray gun straight onto a body or dress form. The solvent then evaporates, and the fibers bond, forming a snug garment. By using different types of fibers, the company can produce different types and textures of fabrics, all of which can hold a wide range of colors and even scents—if that's to your fancy. And when your clothes are sprayed straight on your skin, there's one issue you eliminate altogether: the problem of fit.

You might argue that you don't need aerosol clothes and that you're getting along quite well with your cottons and woolens—and you might be right. But Fabrican has a lot of other—more practical—uses for its technology. The company is exploring medical applications for spray-on surfaces, including new types of bandages, dressings, and skin coverings. It might also be possible to infuse fabrics with antibiotic or other medications, making steady, slow-release dosing easier. Sprayable fabric could be used as well for a range of hygiene or cleaning products, such as inexpensive diapers, paper towels, and sanitary wipes. The company also sees uses in the automotive industry. Spray-on fenders are not likely, but spray-on upholstery—particularly for taxis and other fleet vehicles that get a lot of use and sustain a lot of wear—could be. There might even be uses in interior design: Spray-on wall coverings could replace paint or wallpaper.

It's on the catwalk, however, that Fabrican is making the biggest splash. The first runway show of spray-on clothes took place in the fall of 2010. As with all events in the fashion world, it's too early to know whether the new style will stick.

FROM TOP: GENE KEIGEL; IAN COLE; FABRICAN SPRAY-ON: DR. MAÑUEL TORRES

Replacing Real Legs With eLegs

For paraplegic patients, the ability to stand—not to mention take a few steps—under their own power is a cruelly unattainable goal. Or at least it has been. The makers of eLegs, an innovative exoskeleton, are hoping to change that, one step at a time. The robotic prosthetic legs use artificial intelligence to read the wearer's arm gestures via a set of crutches, and use that information to simulate a natural human gait, effectively carrying the operator along. It's the first such device to do so without a tether, and it was inspired by military exoskeletons—like the Raytheon Iron Man suit (see page 38)—that soldiers strap on to lift heavy packs.

A device like eLegs has been a long time coming and fills an enormous need. About 6 million people in the U.S. suffer from some form of paralysis, a condition that not only leaves them confined, but also contributes to a host of other immobility-related ills like poor circulation, obesity, and pressure sores. The system is designed to handle multiple terrains, which means it need not be confined to indoor use, and it operates comparatively silently. Speed can be adjusted depending on the ability of the wearer, but walking speeds of 2 mph–not a jog, but not a creep either—are possible.

Not everyone is right for a set of eLegs; overall strength, fitness, and balance are important. What's more, the current design can handle body weights of no more than 220 pounds, and the user's height must be no less than 5-foot-2, and no more than 6-foot-4. Still, that covers the large majority of people. Velcro fasteners allow the legs to be put on relatively easily, and they can be worn over clothes. They're not light—the entire system weighs 45 pounds—but once they're in place, they carry their own weight. Even in the best case, the device takes some getting used to. For that reason it will initially be available only at rehabilitation centers for use with a trained physical therapist, but it may hit the home market by 2013.

45 POUNDS
WEIGHT OF EXOSKELETON

2 MPG
MAXIMUM SPEED

6+ HOURS
BATTERY LIFE

MOTIONS: Walk in a straight line, stand from a sitting position, stand for an extended period, and sit down from a standing position

WHO CAN USE: Those who can self-transfer from a wheelchair to a chair and are between 5-foot-2 and 6-foot-4 and weigh 220 pounds or less

FROM TOP: JUNG YEON-JE/AFP/GETTY IMAGES; BARTHOLOMEW COOKE (7)

Charging Up While You Ride

Electric buses in Seoul, South Korea, may never need to stop for a plug-in. That's because the Korea Advanced Institute of Technology is experimenting with embedding electric strips in roadbeds that magnetically transfer energy to battery-powered vehicles above. A prototype at an amusement park in Gwacheon is the first in the world like it, and researchers say the technology could enable all electric vehicles to operate with one-fifth the battery size and at a third of the cost.

The system is designed both efficiently and cleverly. The charging rails embedded in the roads are laid in long, noncontiguous stretches, mostly in bus lanes and at the approaches to intersections. Why intersections? Because vehicles must slow down there and are thus in the vicinity of the rails longer. The rail and the bus do not have to be in physical contact with each other as trams or trolleys linked to overhead lines are. Rather, a process called inductive charging lets magnetic fields do the work. Every time a vehicle gets close enough, its battery gets a boost of juice—what the engineers call microcharges. Accumulate enough of them over the day and you never run out of power. That sidesteps the biggest obstacle to the large-scale adoption of electric vehicles: the limited battery life that requires repeated plug-ins or an onboard gas engine to complement the electric power.

In a perfect and very green world, all cities would build roads with magnetic rails installed, but not only are no cities perfect, plenty of them are broke. It costs about $353,000 per kilometer of road ($565,000 per mile) to lay the rails, and that's exclusive of the cost of energy. Still, environmentalists have increasingly come to accept that there will never be any one innovation that solves the world's energy woes, but rather a great many of them working together. Magnet recharging looks like a promising part of that mix.

Iron Man Is Real—And He's You

If you're a kid, here's one more reason to be a geek: When you grow up, you get to build indescribably cool stuff. Take the XOS 2, developed by Salt Lake City–based Raytheon Sarcos. An honest-to-goodness Iron Man suit, the XOS 2 allows even its least muscular wearer to lift 200-pound weights without breaking a sweat and, as seen in demonstration videos that have gone viral, punch through four one-inch slabs of wood stacked together—something an unsuited person would ordinarily be at pains even to saw through. Raytheon hopes to roll out the XOS 2 to the military, allowing soldiers in theaters of operation to lift heavy ordnance or other equipment with ease.

Iron Man suits are envisioned in two different models—and both are at different stages of development. The combat version, which would be used in the field and help soldiers lighten the loads they often have to carry, is not nearly ready for prime time, since it would have to be energy independent—able to operate untethered on an internal power source far from any generator or outlet. As with electric cars, developing a power pack with that kind of portability has proven challenging. The first iteration of the suit was a power hog, but later versions have doubled the system's efficiency. That's good, but still not good enough for routine use. Raytheon has had a little more success shaving weight from the suit—important since soldiers need to be able to move in it comfortably and safely.

The logistics version of the Iron Man system is proving easier to perfect, since it would operate behind the lines at depots or military construction sites, where the power is more predictable and the idea of working with what amounts to an extension cord running from your suit is not quite so hard to imagine.

Other challenges remain as well. The suit must be fully weatherproofed so it doesn't short out in wet environments or overheat in desert settings. What's more, as robot spacecraft working on the even less forgiving surfaces of Mars and the moon have taught us, sand in the gears can also bollix things up fairly completely. Joints must thus be well protected and tight-fitting to eliminate the risk of any stray grains. Even before the military gets its first Iron Man suits, it's thus likely that the hardware will be turning up in industrial settings—shipyards and factories, where workers grapple with heavy loads all the time and would welcome a little metallic muscle.

17:1

THE RATIO BETWEEN ACTUAL AND PERCEIVED WEIGHT. WITH THE XOS 2 SUIT, A 200-POUND WEIGHT FEELS LIKE ONLY 12 POUNDS.

Baby, You Can't Drive This Car

Is it an auto-automobile? An automaton? Whatever you call it, Google's new driverless Prius has driven itself 140,000 miles without an unscheduled meeting with a light pole. Other geek squads have been running driverless vehicles in the California desert for years, partly at the behest of the U.S. Department of Defense, which has been interested in the technology for a long time. But only Google can rev the petabyte-sucking mapping technology that guides its car along busy streets and highways.

It takes a lot of hardware and software to replace the eyes and brain of a human being. The driverless car is loaded with Google Street View and an artificial-intelligence system, as well as videocameras, radar, and a light-sensing distance and ranging system known as LIDAR. All that technology meshes seamlessly: The Google car has safely navigated San Francisco's serpentine Lombard Street—no one's idea of an easy drive—as well as the Golden Gate Bridge and the Pacific Coast Highway.

The goal of the new system is safety—an admirable one, given the world's million-plus auto fatalities each year. Driverless technology is logical and efficient, and in the near future it could transform your commute into stress-free transport on a motorized sofa. The sad part for road hogs: If Google is truly successful, you may never get to flip the bird at another driver again.

GOOGLE (2)

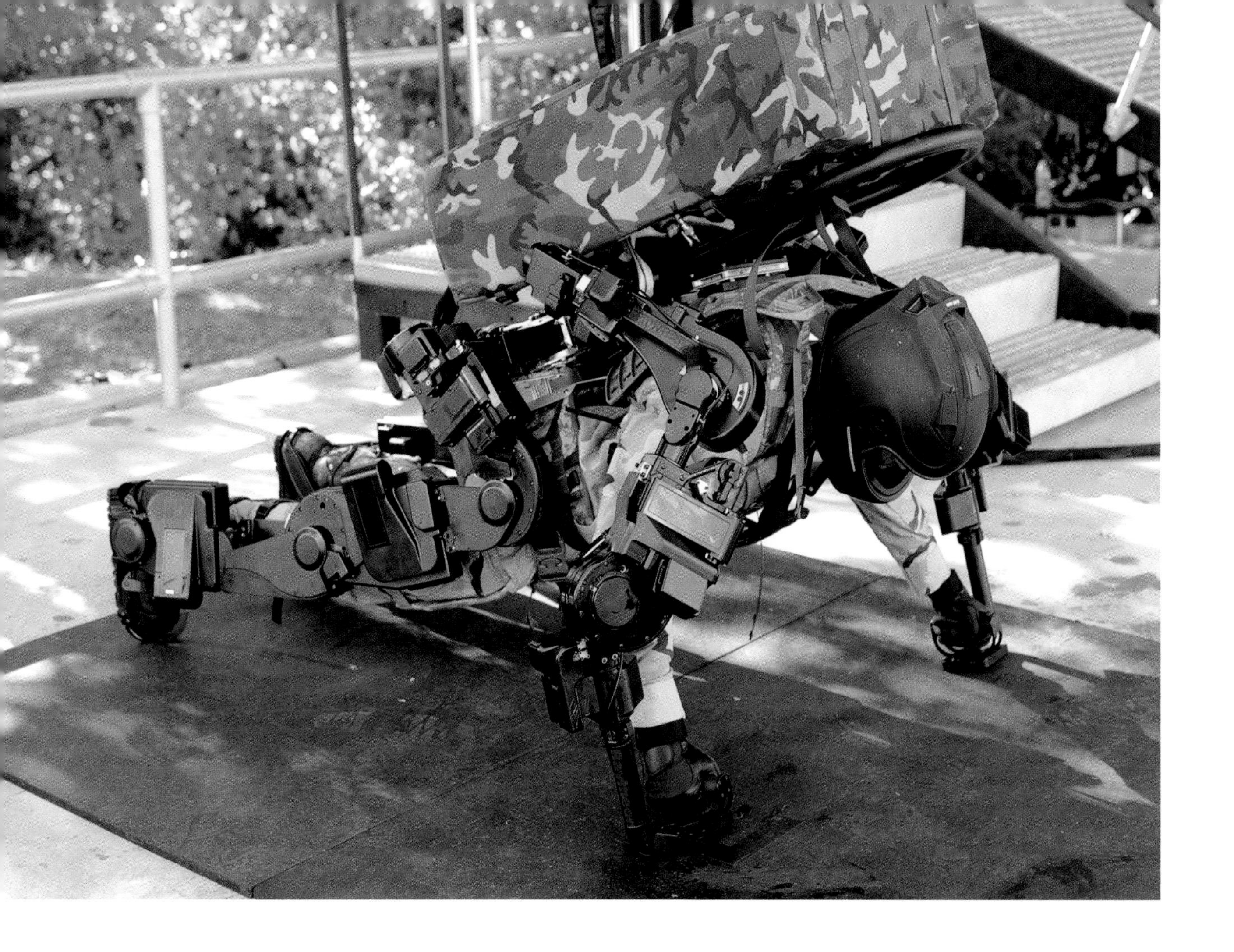

IT JUST AIN'T SO...

WHA' HAPPENED? *Iranian President Ahmadinejad's centrifuges*

Breaking the Firewalls

It's not easy to feel secure in an armed and dangerous world. If it's not nukes, it's terrorism; if it's not terrorism, it's bioweapons. Even our home computers are always in danger of crashing and burning, thanks to some malicious virus. If there's one consolation, it's that military computers—particularly the ones that control our weapons systems—are surrounded by so many layers of firewall that no hacker could ever get through. Um, think again. If the lesson Iran is currently learning is any indication, no weapons system is safe anymore.

Iran, certainly, is one of the world's bad actors—and its current push to develop nuclear weapons makes it worse than ever. It was thus good news to learn that the country's nuclear labs had encountered technical problems lately, with many of the centrifuges needed to enrich nuclear fuel suddenly spinning out of control and destroying themselves. More curiously, even as the centrifuges were going haywire, the plant's sensing equipment continued to indicate that everything was running normally, preventing emergency shutdown systems from kicking in. All that, analysts now agree, was the result of a computer worm called Stuxnet, jointly designed by the U.S. and Israel. Neither country will acknowledge it, but they probably don't need to. Nobody will mourn anything that gets in the way of an Iranian bomb, but if the bad guys are vulnerable, the good guys could one day be as well. Even as we raise our offensive game, we need to play tougher defense too.

FROM TOP: RAYTHEON; GETTY IMAGES

Archaeology

THE PRE–*T. REX* ■ WHAT KILLED KING TUT ■ THE HOBBITS OF FLORES
EGYPT'S ANCIENT BAKERY ■ UNRAVELING NEANDERTHAL DNA ■ HOW FEATHERS CAME TO BE
THE SPLENDOR OF EL MIRADOR ■ THE FIRST GREAT CITY ■ THE 2 MILLION-YEAR-OLD BOY

A Long-Lost Relative

The oldest hominid skeleton ever discovered offers unexpected clues to what our even more ancient ancestors might have looked like.

By Michael D. Lemonick and Andrea Dorfman

Figuring out the story of human origins is like assembling a huge, complicated jigsaw puzzle that has lost most of its pieces. Many will never be found, and those that do turn up are sometimes hard to place. Every so often, though, fossil hunters stumble upon a discovery that fills in a big chunk of the puzzle all at once—and simultaneously reshapes the very picture they thought they were building.

The path of just such a discovery began in November 1994 with the unearthing of two pieces of bone from the palm of a hominid hand in the dusty Middle Awash region of Ethiopia. Within weeks, more than 100 additional bone fragments were found during an intensive search and reconstruction effort that would go on for the next 15 years and culminate in a key piece of evolutionary evidence revealed in 2009: the 4.4 million-year-old skeleton of a likely human ancestor known as *Ardipithecus ramidus.*

In a series of landmark studies—11 papers by 47 authors from 10 countries—researchers unveiled "Ardi," a 125-piece hominid skeleton that is 1.2 million years older than the celebrated Lucy *(Australopithecus afarensis)* and by far the oldest one ever found. Tim White of the University of California at Berkeley, a co-leader of the Middle Awash research team that discovered and studied the new fossils, says, "To understand the biology, the parts you really want are the skull and teeth, the pelvis, the limbs, and the hands and the feet. And we have all of them."

That is the beauty of Ardi—good bones. The completeness of Ardi's remains, as well as the more than 150,000 plant and animal fossils collected from surrounding sediments of the same period, has generated an unprecedented amount of intelligence

TIM D. WHITE

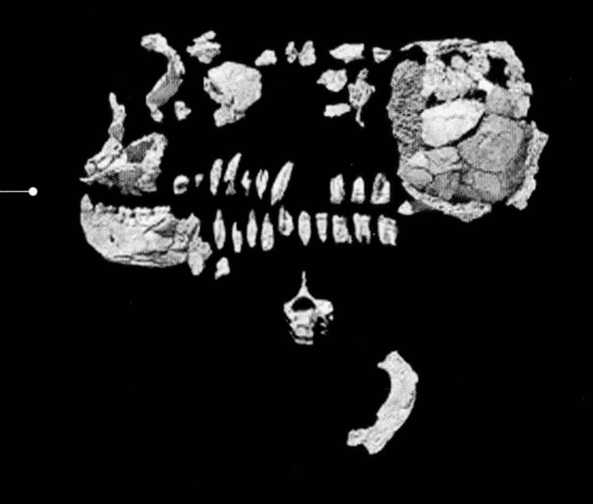

TEETH
Their size, shape, structure, and enamel composition indicate that Ardi was omnivorous. Males of her species lacked the daggerlike fangs of gorillas and chimps, suggesting that *Ardipithecus* didn't fight over mates.

PELVIS
The broad upper portion allowed the lumbar (lower back) vertebrae to curve inward, essential for upright walking. The apelike lower pelvis anchored powerful hamstring muscles used for climbing.

FEET
Unlike any later hominid, Ardi had an opposable, grasping big toe that aided in climbing. The rest of her flat foot was rigid enough to act as a propulsive lever when she walked on two legs. Her gait could be somewhat clumsy, and if she ran, she would tire quickly.

HANDS
Ardi didn't swing through trees much, but her long, dexterous fingers and flexible palms were ideal for grasping. Her wrists were equally flexible, enabling her to bend her hands back and "palm-walk" along branches, just as extinct apes did.

VITAL STATISTICS
Female, most likely a young adult; 47 inches tall and 110 pounds

about one of our earliest potential forebears. The skeleton allows scientists to compare *Ardipithecus* with Lucy's genus, *Australopithecus*, its probable descendant. Perhaps most important, Ardi provides clues to what the last common ancestor shared by humans and chimps may have looked like before their lineages diverged about 7 million years ago.

Ardi is the earliest and best-documented descendant of that common ancestor. But despite being "so close to the split," says White, the surprising thing is that she bears little resemblance to chimpanzees. The elusive common ancestor's bones have never been found, but scientists, working from the evidence available, especially analyses of *Australopithecus* and modern African apes, envisioned a great-great-grandpa who looked most nearly like a knuckle-walking, tree-swinging ape.

But Ardi is "not chimplike," according to White, which means that the last common ancestor probably wasn't either. "This skeleton flips our understanding of human evolution," says Kent State University anthropologist C. Owen Lovejoy, a member of the Middle Awash team. "Humans are not merely a slight modification of chimps."

What does that mean? Based on Ardi's anatomy, it appears that chimpanzees may actually have evolved more than humans—in the scientific sense of having changed more over the past 7 million years or so. That's not to say that Ardi was more humanlike than chimplike. White describes her as an "interesting mosaic" with certain uniquely human characteristics: bipedalism, for one. Ardi stood 47 inches tall and weighed about 110 pounds, making her roughly twice as heavy as Lucy. The structure of Ardi's upper pelvis, leg bones, and feet indicates she walked upright on the ground, while still retaining the ability to climb. Her foot had an opposable big toe for grasping tree limbs but lacked the flexibility that apes use to grab and scale tree trunks and vines ("Gorilla and chimp feet are almost like hands," says Lovejoy); nor did it have the arch that allowed *Australopithecus* and *Homo* to walk without lurching from side to side. Ardi had a dexterous hand, more maneuverable than a chimp's, that made her better at catching things on the ground and carrying things while walking on two legs. Her wrist, hands, and shoulders show that she wasn't a knuckle walker and didn't spend much time hanging ape-style in trees. Rather, she moved along branches using a primitive method of palm walking typical of extinct apes. "[Ardi] has features that are intermediate between the last common ancestor and australopithecines," says Penn State paleoanthropologist Alan Walker, who was not involved in the discovery.

Scientists know this because they've studied not only Ardi's fossils but also 110 other remnants they uncovered, which belonged to at least 35 *Ar. ramidus* individuals. Combining those bones with the thousands of plant and animal fossils from the site yielded a remarkably clear picture of the habitat Ardi roamed some 200,000 generations ago. It was a grassy woodland with patches of denser forest and freshwater springs. Colobus monkeys chattered in the trees, while baboons, elephants, spiral-horned antelopes, and hyenas roamed the terrain. There was an assortment of bats and at least 29 species of birds, including peacocks, doves, lovebirds, swifts, and owls. Buried in the Ethiopian sediments were hackberry seeds, fossilized palm wood, and traces of pollen from fig trees.

This tableau demolishes one aspect of what had been conventional evolutionary wisdom. Paleoanthropologists once thought that what got our ancestors walking on two legs in the first place was a change in climate that transformed the African forest into savanna. In such an environment, goes the reasoning, upright-standing primates would

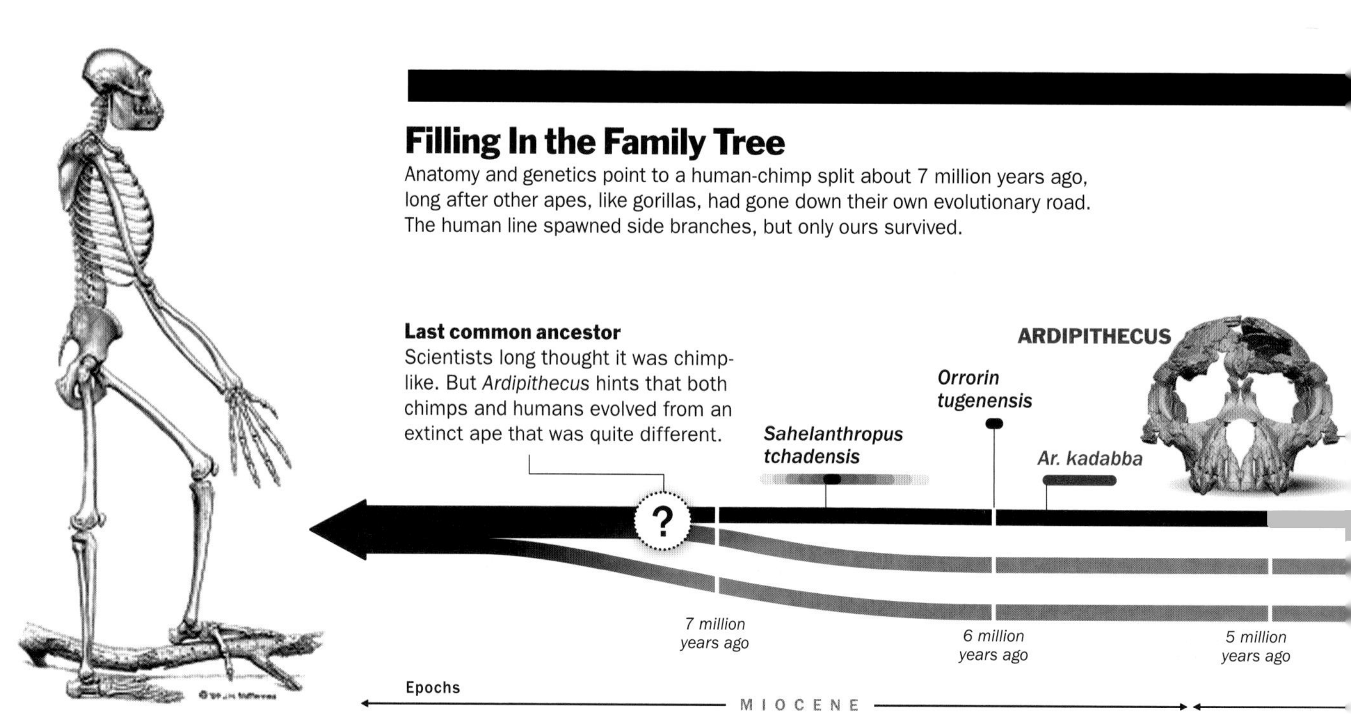

PRESERVATION LIQUID *Adhesive helps keep fossils from crumbling.*

"It's clear that humans are not merely a slight modification of chimps, despite their genomic similarity."

—C. OWEN LOVEJOY, ANTHROPOLOGIST

have had an advantage over knuckle walkers because they could see over tall grasses to find food and avoid predators. The fact that Lucy's species sometimes lived in a more wooded environment began to undermine that theory. The fact that Ardi walked upright in a similar environment many hundreds of thousands of years earlier makes it clear that there must have been another reason.

No one knows what that reason was, but a theory about Ardi's social behavior may hold a clue. Lovejoy thinks *Ar. ramidus* had a social system found in no other primates but humans. Among gorillas and chimps, males viciously fight other males for the attention of females. But among *Ardipithecus,* says Lovejoy, males may have abandoned such competition, opting instead to pair-bond with females and stay together to rear offspring (though not necessarily monogamously or for life). The evidence of this harmonious existence comes from *Ardipithecus* teeth: Its canine teeth are stubby compared with the sharp, daggerlike upper fangs that male chimps and gorillas use to do battle. "The male canine tooth is no longer weaponry," says Lovejoy.

That suggests that females mated preferentially with smaller-fanged males. In order for females to have had so much power, Lovejoy argues, *Ar. ramidus* must have developed a social system in which males were cooperative. Males probably helped females and their own offspring by foraging for and sharing food, for example—a change in behavior that could help explain why bipedality arose. Carrying food is difficult in the woods, after all, if you can't free up your forelimbs by walking erect.

Deducing such details of social behavior is, admittedly, speculative. One problem is that some portions of Ardi's skeleton were found crushed nearly to smithereens and needed extensive digital reconstruction. "Tim [White] showed me pictures of the pelvis in the ground, and it looked like an Irish stew," says Walker. Indeed, looking at the evidence, different paleoanthropologists may have different interpretations of Ardi's biomechanics and behavior.

But the extraordinary number and variety of fossils uncovered in the place Ardi came to rest show that scientists are arguing over real evidence, not the usual single tooth here or bit of foot bone there. "When we started our work [in the Middle Awash]," says White, "the human fossil record went back to about 3.7 million years." Now scientists have a trove of information from an era some 700,000 years closer to the dawn of the human lineage. "This isn't just a skeleton," he says. "We've been able to put together a fantastic, high-resolution snapshot of a period that was a blank." The search for more pieces continues, but the outlines of the puzzle, at least, are coming into focus.

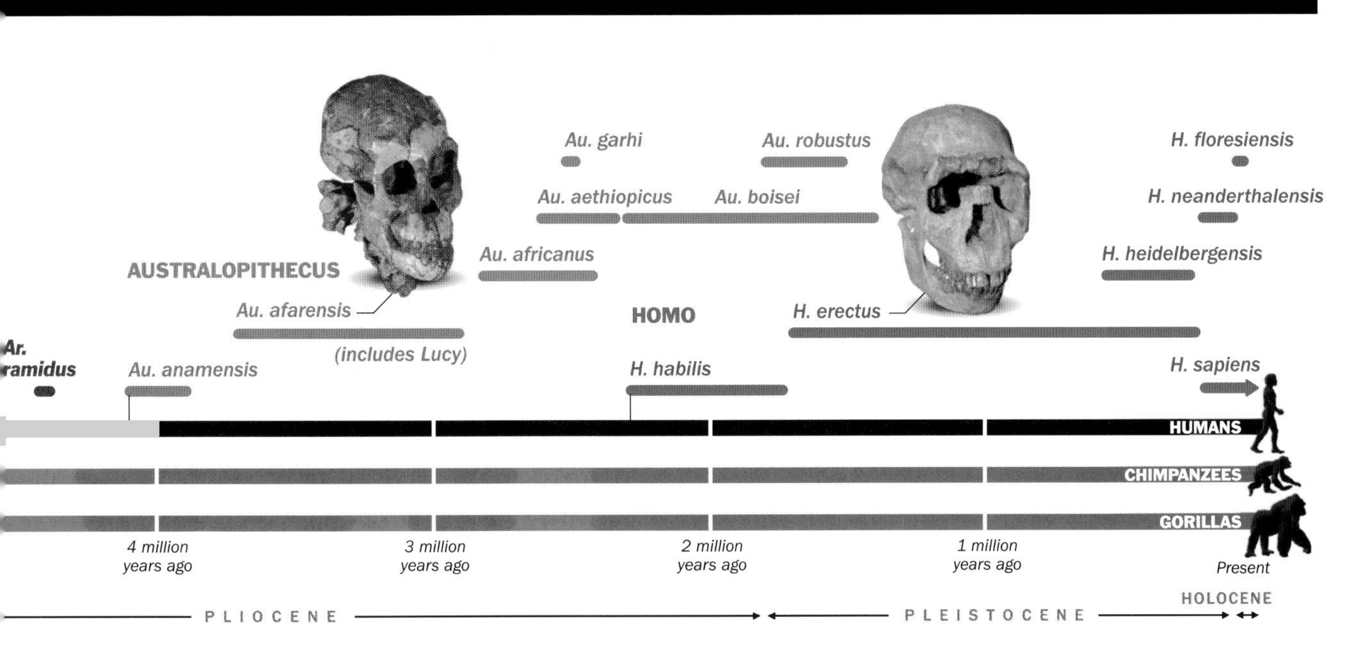

2,000 FEET

THE DEPTH OF THE LAVA BED IN WHICH SOME OF THE FOSSILS OF THE ANCIENT ANDES ARE BURIED

LITTLE KILLER *A model shows the lethal teeth that would later define the* T. Rex.

Say Hello to the Pre-*Rex*

A *T. Rex* might have scared the daylights out of you had you been around 65 million years ago and happened to run into one. But its great-great-granddad, the recently discovered dawn runner, which lived 230 million years back? Not so much. Despite its speed, its teeth, and its predatory ways, the dawn runner—officially known as the *Eodromaeus*—measured barely 48 inches from its nose to the end of its tail and weighed in at no more than 15 pounds. Its significance is not so much its fearsomeness but its bloodlines—and what they tell us about the family trees of all other dinosaurs as well.

First described in a paper in the journal *Science* in 2011, the dawn runner was discovered by a team of American and Argentine paleontologists in what is now the foothills of the Andes Mountains and was once the southwest region of the supercontinent known as Pangaea. The trove of remains discovered along with *Eodromaeus* reveals an era in which dinosaurs were actually in the minority, outnumbered by smaller reptiles. And like the dawn runner, the other fossils appear to be early representatives of much later species, which took far longer than paleontologists realized to emerge. This suggests a new scenario in which dinosaurs didn't storm into dominance all at once, perhaps as a result of some global cataclysm that wiped out other species. Rather, they crept slowly into control, eventually ruling the world for 163 million years until their own era came to an end, making way for mammals—including us.

King Tut: It Was Malaria, Not Murder, That Did Him In

The *Journal of the American Medical Association* is not the place murder mysteries are usually reported. But when the weapon used in the killing turns out to be malaria and the victim was Tutankhamen, the boy pharaoh better known as King Tut, the journal makes an exception.

Popular theory had long held that Tut was murdered, not implausible in an ancient palace court, especially when a ruler died so young. To settle the question, Egyptian officials called in German geneticist Carsten Pusch to sample Tut's DNA and see whether anything in his genes might have claimed him first. Among the results: He did not have any of the diseases that experts have speculated about, including Marfan syndrome, Wilson-Turner X-linked mental retardation syndrome, and androgen-insensitivity syndrome. What probably did Tut in, says Pusch, was an immune system that was badly compromised by a particularly virulent strain of malaria combined with a degenerative bone disease that had already left him weak. The DNA also showed that feminized artistic depictions of Tutankhamen and Akhenaten, Tut's father and predecessor, with breasts were only that: There was no evidence of hormonal imbalances that could have resulted in the real thing.

ANCIENT REDOUBT *One million years ago the hobbits arrived on Flores. In 2003 their remains were first discovered in this cave.*

CLOCKWISE FROM RIGHT: ROSINO/FLICKR; ETHAN MILLER/GETTY IMAGES; MIKE HETTWER COURTESY OF PROJECT EXPLORATION (2)

The Hobbit Bones of Flores

Anthropologists did not know quite what they had when they first stumbled on the intriguing remains of undersized humans—just over 3 feet tall—on the Indonesian island of Flores in 2003. The bones were old—dating from 95,000 years ago—and they were clearly adult. Ultimately, they were determined to be a species, officially named *Homo floresiensis*—and unofficially called hobbits. Their diminutive stature was attributed to what's known as the island effect, a long-term adaptation to confined living and scarce resources driven by the simple metabolic equation that small bodies need less food. The fact that pygmy elephants also lived on the island is one sign that the theory is correct.

Fascination with the hobbits has never waned, and recent finds of stone tools make it clear that their ancestors were living on the island far longer than the hobbits themselves were—perhaps 1 million years ago. Anthropologist William Jungers of Stony Brook University in New York argues that those pre-hobbits traveled a long and dangerous route to get to Flores, island-hopping from mainland Asia, at last settling down, and presumably only then adjusting their size to their surroundings over the course of many generations. Whatever their life was like, it apparently served them well, since they survived on Flores until 17,000 years ago.

Research on the island is ongoing, in part because not all anthropologists agree on the significance of the hobbit remains. Shortly after the discovery of new stone tools was announced in March 2010, anthropologists holding their annual meeting in Albuquerque debated the very legitimacy of the species, some arguing that they are just an early version of modern pygmies and others attributing their small size to a developmental disorder of the skeleton.

3'6"

HEIGHT OF THE FIRST HOBBIT FOUND. ITS WEIGHT—ESTIMATED BY ITS HEIGHT AND ITS OVERALL BONE STRUCTURE—WAS PROBABLY 65 POUNDS.

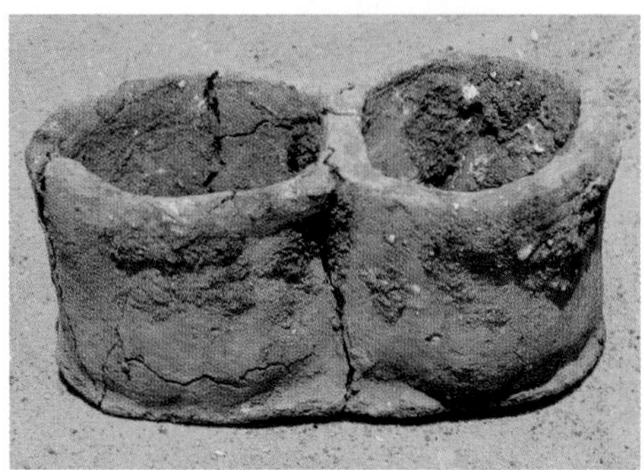

REMAINS OF THE OASIS *(clockwise from bottom) A cupcake-like bread mold, a Nubian bowl, and a stone seal*

Egypt's Ancient Bakery Shop

The Theban Desert in Egypt can be a punishing place. A blistering expanse of limestone and sand, it ought to be home to nothing at all. And it is, for the most part. But the Kharga Oasis is—or at least once was—another matter entirely. In a major find announced in 2010, a husband and wife team from Yale University—John and Deborah Darnell—discovered the remains of an agricultural, mercantile, and military center that thrived in Kharga roughly 3,500 years ago.

Following the ancient caravan routes, the Darnells came first upon a large cache of pottery—so large that it had to have been produced locally. Later they found the remains of workshops, silos, stockrooms, and what appeared to be an administrative building. Most intriguing, they found a huge bakery complex—so large it was probably used to feed an army. The site has since been dubbed Umm Mawagir, or "mother of bread molds." The commercial and military uses of the oasis fit nicely with that era in Egyptian history, when the country was under siege by its neighbors and the central desert was the safest place to build—and slowly expand—what would eventually become an empire.

A HALF TON OF BAKERY ARTIFACTS HAVE BEEN DISCOVERED IN UMM MAWAGIR—EVIDENCE OF A MILITARY GARRISON.

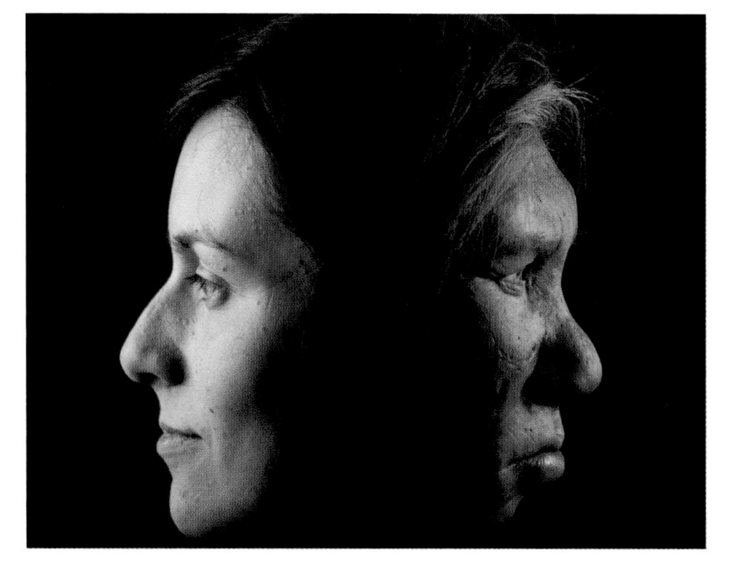

CLOSE KIN *Analyses of both human and Neanderthal DNA suggest that the two species may have interbred.*

Neanderthal and Humans: Was There Genetic Canoodling?

A decade after scientists first cracked the human genome, researchers announced that they had done the same for Neanderthals, the species of hominid that existed from roughly 400,000 to 30,000 years ago, when their closest relatives, early modern humans, may have driven them to extinction. Led by ancient-DNA expert Svante Pääbo of Germany's Max Planck Institute, scientists reconstructed about 60% of the Neanderthal genome by analyzing tiny chains of ancient DNA extracted from the bone fragments of three female Neanderthals excavated in the late 1970s and early '80s from a cave in Croatia. The bones are 38,000 to 44,000 years old.

The biggest surprise: The genetic information suggests that at some point after early modern humans migrated out of Africa, they mingled and mated with Neanderthals, possibly in the Middle East or North Africa as much as 80,000 years ago. If that is the case, it occurred significantly earlier than scientists who support the interbreeding hypothesis would have expected. Comparisons with DNA from modern humans even show that some Neanderthal DNA has survived to the present.

FROM TOP: THEBAN DESERT ROAD SURVEY (3); JOE MCNALLY

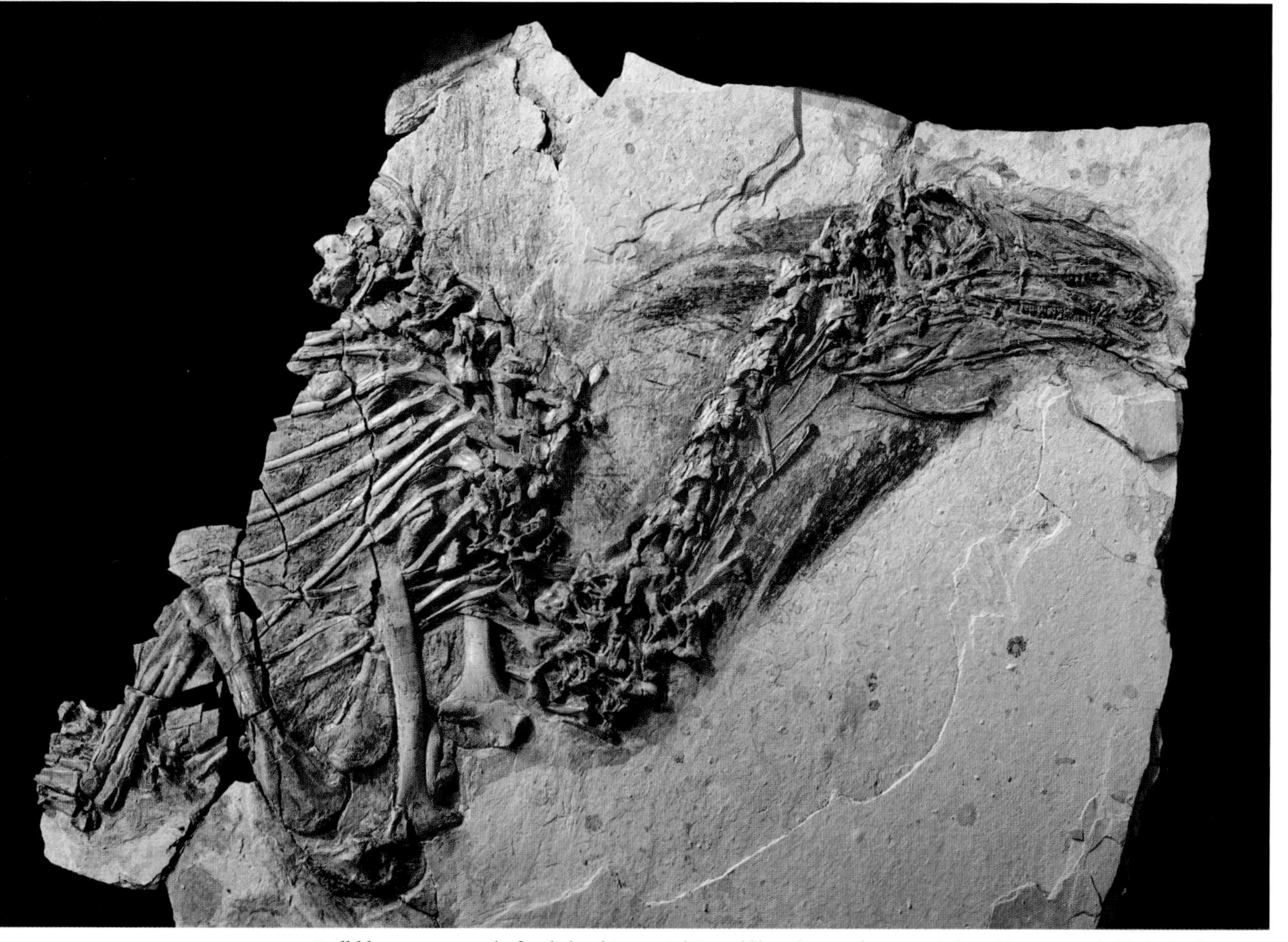

PRETTY PLUMAGE *Quill-like structures on the fossil's head (upper right) would have been useless to retain heat. They were probably for display.*

FROM TOP: ROBERT CLARK/INSTITUTE; JULIUS T. CSOTONY/PHOTO RESEARCHERS

Dinos, Birds, and the Evolution of Feathers

It's hard to draw a straight line between the dinosaurs of the late Cretaceous and the pigeons of a modern Chicago or Manhattan, but as paleontologists have known for a good while now, a very direct connection exists. Dinosaurs evolved into birds, as overwhelming evidence of their skeletal remains suggests. What's always been less clear is just why feathers evolved—if they were always intended for flight or were originally used for warmth, and what their color scheme was. Now we have a much better idea.

British, Irish, and Chinese paleontologists reported in 2010 on a study of fossils found in northeastern China. The remains unearthed —from the theropod *Sinosauropteryx* and the primitive bird species *Confuciusornis*—were special not just because the feathers survived but because the color-bearing organelles that give them their distinctive hues were also intact. Known as melanosomes, they are responsible for producing shades including gray, orange, and brown. The feathers in which they were found were not fully feathers, but a sort of pre-feather bristles. In the case of the *Sinosauropteryx,* they formed a pattern of orange and white tail rings; in the case of the *Confuciusornis* they created a patchwork pattern.

The fact that the feathery structures were so crude and were present in the absence of wings suggests that they were indeed there for insulation and that their elaborate colors were used for display. Only later did they evolve for flight. Other work by Beijing-based paleontologist Xing Xu found more-elaborate quill-like feathers on an adult *Similicaudipteryx* fossil, from a time 25 million years after the first feather-like structures emerged in earlier species.

MILLENNIAL ART *Archaeologist Richard Hanson shows off a 2,300-year-old frieze.*

The Most Amazing Place You Never Heard Of

Quick—where could you find the world's largest ancient pyramid? Wrong. (We're assuming you said Egypt.) Okay, where could you find it in the Western Hemisphere? Wrong again. (We're assuming you said Mexico or Machu Picchu.) Where you'd actually have to go to find the king of pyramids is El Mirador, the earliest and largest Mayan city-state ever discovered, located in Guatemala. If you've never heard of El Mirador, you're not alone, but the vast sprawl of ancient ruins have been the ongoing object of archaeological fascination and study since the 1970s.

The El Mirador city-state flourished up to 2,300 years ago, but its largest structures date to what is known as the Late Preclassic period, from 350 B.C. to A.D. 150. It was in that era that the La Danta temple was built, topping out at a height of 230 feet, with a record-setting volume of 99 million cubic feet. But El Mirador had its cozier side too, with clusters of residences similar to modern-day housing developments and major avenues tying them all together into a dense web of communities and commerce. Some of the forms and techniques used in constructing the buildings are unseen anywhere else in the world.

With its relatively short history as an archaeological site—the first major excavation was begun in 1978—El Mirador has not yet revealed even a fraction of its secrets. Currently scientists from 34 universities and other research centers around the world are working cooperatively at the site. One of the great mysteries they have yet to answer: What caused the collapse of the jungle metropolis in A.D. 150, and what clues to that disaster can be found in the ruins?

100,000

PEOPLE PROBABLY LIVED IN EL MIRADOR AT ITS PEAK. A SMALLER POPULATION WOULD HAVE BEEN UNABLE TO BUILD SO SPRAWLING A PLACE.

The First of the World's Greatest Cities

One of the most exciting places on earth, provided you lived from 5,500 to 4,000 years ago, had to be the ancient city of Tell Zeidan, located in northern Syria. Long eclipsed by the region's other great archaeological sites in Israel, Egypt, Iraq, and elsewhere, Tell Zeidan is now getting its overdue due.

In 2008 and 2009, American and Syrian archaeologists kicked off a program of exploration and excavation that could easily continue for a generation. Even the early finds make it clear why. Investigators have discovered a complex of eight kilns that suggest a population that had the ability not just to fire pottery, but to do so in an organized and industrialized way. Artifacts used in copper smelting have also been found around the site—high-tech stuff in that long-ago era. Remains of homes and fortifications also point to an established—and well-protected—community that both traded in goods and sought to guard its wealth. A carved stamp of a deer may have been a commercial tool, perhaps used to mark goods for sale or seal a package before shipment.

CULTURAL BITS *Painted ceramics from Tell Zeidan show a style brought from Mesopotamia.*

Other such instruments were found elsewhere in Tell Zeidan, a sign of a hierarchical culture in which some people had already risen to positions of commercial influence, regulating commerce and trade. This early stratification of the population—what we would call a class system—is a feature of all cities, but this could be the first clear sign of it.

FROM TOP: EDUARDO GONZALES/AFP/GETTY IMAGES; COURTESY GIL STEIN

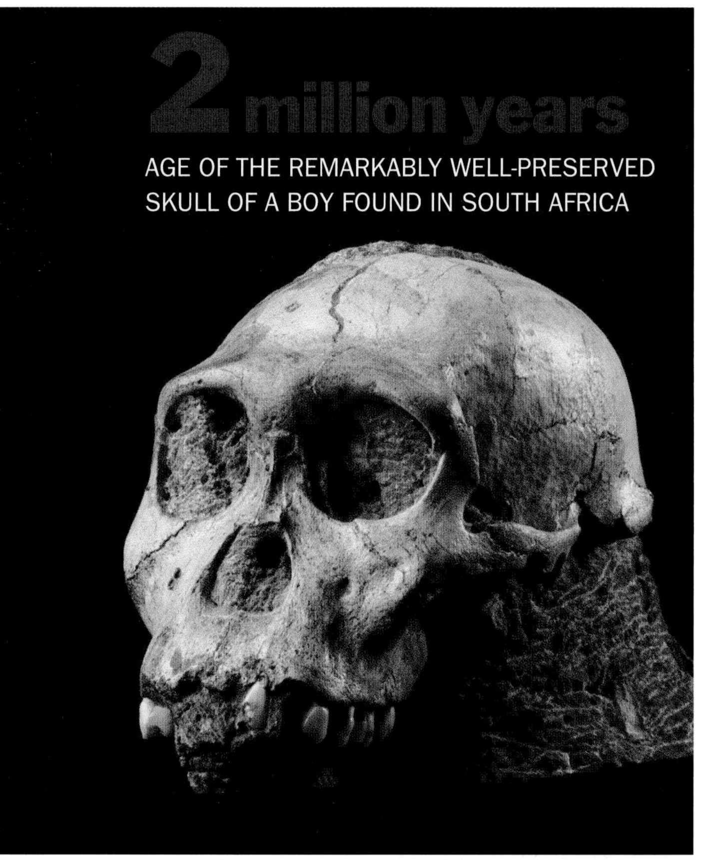

MASH-UP *This ancient boy had a small brain but modern features.*

South Africa's Mystery Hominids

We can never know much about the woman and boy who were entombed in an avalanche of sediment in South Africa's Malapa cave some 2 million years ago, but spare them a thought, since they might have been kin. Described in the journal *Science* in 2010, the fossils could fill an important spot in the evolutionary arc of humans, since there's little skeletal evidence of what was going on at that particular moment in our history. But not all paleontologists agree about the significance of the new species, dubbed *Australopithecus sediba,* and some dismiss it as an evolutionary dead end.

Part of the debate is due to the fossils' mix of ancient and relatively modern features. The arms are long and apelike, suggesting *Au. sediba* was a good tree climber. The hands are apelike in their curvature but surprisingly compact, like a more modern hominid's. And the longer legs and short pelvis are decidedly modern.

The boy's skull is a similar mash-up of older and newer features. (The woman's skull is missing.) It is quite small and would have held a brain hardly bigger than that of *Australopithecus afarensis,* the 3.2 million-year-old species whose best-known example is the celebrated Lucy. But many of the boy's facial features resemble those of later *Homo erectus* and *Homo sapiens.* Ancestor of humans or not, the pair are having their time in the sun—2,000 millennia after their death.

FROM TOP: BRETT ELOFF, COURTESY LEE BERGER AND THE UNIVERSITY OF THE WITWATERSRAND; PAUL GAUGUIN/GETTY IMAGES

IT JUST AIN'T SO ...

Who Were the First Polynesians?

The lure of Polynesia has long been powerful. French painter Paul Gauguin couldn't resist it, nor could the mutinous sailors of the *Bounty*—nor can modern honeymooners from around the world. The question of who first succumbed to the islands' charms has long been considered a settled matter in the field of archaeology. It was voyagers from Taiwan 3,000 years ago who initially laid claim to Polynesia, bringing with them their ancient language, their distinctive brand of ceramics, and their obsidian tools—all of which survive in various forms today and all of which firmly prove the Taiwanese provenance.

Or all of which once seemed to prove it. But scraps of language and bits of tools are nothing compared to the precision of DNA, and a study published by geneticist Martin Richards of the University of Leeds now demolishes the old belief. Richards used samples of what's known as mitochondrial DNA, which, as its name suggests, is the genetic material stored in the cell's mitochondria, a biological marker that can be used to trace maternal descent back thousands of years. Richards' analysis shows that the very first Polynesian settlers descended from Asian mainlanders who had already arrived on the islands near Papua New Guinea a full 6,000 to 8,000 years ago.

Richards does not dismiss the Taiwanese role. He believes that trade between Taiwan and Polynesia is more than sufficient to account for both the linguistic and the artisanal similarities between the two populations. But as for who came first? DNA doesn't lie.

ISLAND COLOR *A Gauguin masterpiece from 1892*

Genetics

BREEDING ANIMALS BACK FROM EXTINCTION ■ CREATING ARTIFICIAL LIFE ■ GENETIC PATTERNS IN AUTISM
GENES AND YOUR WEIGHT ■ REWRITING HISTORY BOOKS WITH GENETICS ■ VAULTS TO SAFEGUARD OUR FOODS
SCARS OF THE HOLOCAUST ■ THE AGING GENE ■ BACTERIA'S SUPERGENE

Why Genes Aren't Destiny

The new field of epigenetics is showing how your environment and your choices can influence your genetic code–and that of your kids.

BY JOHN CLOUD

The remote, snow-swept expanse of northern Sweden is an unlikely place to begin a story about cutting-edge genetic science. The kingdom's northernmost county, Norrbotten, is nearly free of human life; an average of just six people live in each square mile. Yet this tiny population can reveal a lot about how genes work in our everyday lives.

Norrbotten is so isolated that in the 19th century, if the harvest was bad, people starved. When the harvest was good—which happened just as often—they gorged themselves. In the 1980s, Dr. Lars Olov Bygren, a preventive-health specialist who is now at the prestigious Karolinska Institute in Stockholm, began to wonder what long-term effects the feast and famine years might have had on children growing up in Norrbotten in the 19th century—and not just on them but on their kids and grandkids as well. So he drew a random sample of 99 individuals born in the Overkalix parish of Norrbotten in 1905 and used historical records to trace their parents and grandparents back to birth. He also researched agricultural records to identify the years of feast and famine.

Around the time he started collecting his data, Bygren had become fascinated with research showing that conditions in the womb could affect your health not only when you were a fetus but well into adulthood. In 1986, for example, *The Lancet* published the first of two ground-breaking papers showing that if a pregnant woman ate poorly, her child would be at significantly higher-than-average risk for cardiovascular disease as an adult. It was a heretical idea. After all, we have had a long-standing deal with biology: Whatever choices we make during our lives might ruin our short-term memory or make us fat or hasten death, but they won't change our genes—our actual DNA. Which meant

LARS TUNBJORK/VU/AURORA PHOTOS

THREE GENERATIONS *Dr. Lars Olov Bygren, 73, with son Magnus, 41, and grandson Ludvig, 2. Bygren has studied how men and women born in his father's small hometown passed down new traits in just a generation or two.*

that when we had kids of our own, the genetic slate would be wiped clean.

But Bygren and other scientists have now amassed historical evidence suggesting that powerful environmental conditions (near-death from starvation, for instance) can somehow leave an imprint on the genetic material in eggs and sperm. These genetic imprints can short-circuit evolution and pass along new traits in a single generation.

For instance, Bygren's research showed that in Overkalix, boys who enjoyed those rare overabundant winters—kids who went from normal or poor eating to gluttony in a single season—produced sons and grandsons who lived shorter lives. Far shorter: The grandsons of Overkalix boys who had overeaten died an average of six years earlier than the grandsons of those who had endured a poor harvest. Once Bygren and his team controlled for certain socioeconomic variations, the difference in longevity jumped to an astonishing 32 years. To put it simply, the data suggested that a single winter of overeating as a youngster could initiate a biological chain of events that would lead one's grandchildren to die decades earlier than their peers did. How could this be possible?

The answer lies beyond both nature and nurture, and instead with a new science called epigenetics. At its most basic, epigenetics is the study of changes in gene activity that don't involve alterations to the genetic code but still get passed down to at least one successive generation. It is epigenetic marks sitting on top of the actual genome that tell your genes to switch on or off, to speak loudly or whisper.

Epigenetics brings both good news and bad. Bad news first: There's evidence that lifestyle choices like smoking and eating too much can alter the epigenetic marks atop your DNA in ways that cause the genes for obesity to express themselves too strongly and the genes for longevity to express themselves too weakly. What's more, those same bad behaviors can also predispose your kids—before they are even conceived—to disease and early death.

The good news: Scientists are learning to manipulate epigenetic marks in the lab, which means they are developing drugs that treat illness simply by silencing bad genes and jump-starting good ones. In 2004 the Food and Drug Administration (FDA) approved an epigenetic drug for the first time. Azacitidine is used to treat patients with myelodysplastic syndromes (MDS), a group of rare and deadly blood malignancies. Since 2004 the FDA has approved three other epigenetic drugs that are thought to work at least in part by stimulating tumor-suppressor genes. Epigenetics could also help explain certain scientific mysteries that traditional genetics never could: for instance, why one member of a pair of identical twins can develop bipolar disorder or asthma even though the other is fine.

As momentous as epigenetics sounds, the chemistry of at least one of its mechanisms is fairly simple. Darwin taught us that it takes many generations for a genome to evolve, but researchers have found that it takes only the addition of a methyl group to change an epigenome. A methyl group is a basic unit in organic chemistry: one carbon atom attached to three hydrogen atoms. When a methyl group attaches to a specific spot on a gene it can change the gene's expression.

How You Can Change Your Genes

LARS TUNBJORK/VU/AURORA PHOTOS

FOLLOWING THE TRAIL
Bygren with a photo of his father. What Dad ate long ago, and what his own father ate decades earlier, could still be affecting their descendants.

1
THE STRUCTURE OF CELLS

2
THE ROLE OF DNA AND GENES

3
WHERE TO FIND EPIGENETIC MARKS

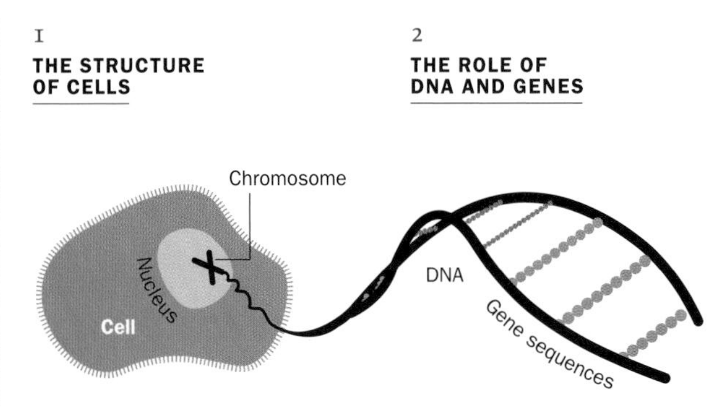

Epigenetic mark

Gene

The human body has trillions of cells, each one with a nucleus, its command center. In each nucleus DNA is tightly coiled around proteins called histones that work as support structures for genes.

Genes contain the codes for cells to produce the various proteins that organisms need to function. Humans have some 25,000 genes. Darwin and his followers taught us that it takes many generations to rewrite this basic genetic code.

Just as genes provide the codes for producing proteins, various chemicals called epigenetic marks sit atop genes and offer basic instructions to them, telling them to switch on or off.

Can epigenetic changes be permanent? Possibly, but it's important to remember that epigenetics doesn't change DNA. Epigenetic changes represent a biological response to an environmental stressor. That response can be inherited through many generations, but if you remove the environmental pressure, the epigenetic marks will eventually fade.

All that explains why the scientific community is so excited. Researchers are quietly acknowledging that we may have too easily dismissed an early naturalist, Jean-Baptiste Lamarck (1744–1829), who argued that evolution could occur within a generation or two. He posited that animals acquired certain traits during their lifetime because of their environment and choices. The most famous Lamarckian example: Giraffes acquired long necks because their recent ancestors had stretched to reach high, nutrient-rich leaves.

In contrast, Darwin argued that evolution works not through the fire of effort but through cold, impartial selection. According to Darwinist thinking, giraffes got their long necks over millennia because genes for long necks very slowly gained advantage. Darwin won the day, and Lamarckian evolution came to be seen as a scientific blunder.

Yet epigenetics is now forcing scientists to reevaluate Lamarck's ideas—and some of those ideas can affect us in small and very personal ways. Studies have suggested that baby lotions containing peanut oil may be partly responsible for the rise in peanut allergies, for example. High maternal anxiety during pregnancy may be associated with a child's later development of asthma; little kids who are kept too clean may be at higher risk for eczema.

How can we harness the power of epigenetics for good? In 2008 the National Institutes of Health (NIH) announced it would pour $190 million into a multilab, nationwide epigenetics initiative. In October 2010, the grant started to pay off. Scientists working jointly at a fledgling, largely Internet-based effort named the San Diego Epigenome Center announced with colleagues from the Salk Institute that they had produced "the first detailed map of the human epigenome."

The human genome contains about 25,000 genes, but the number of epigenetic marks is so large—maybe millions—that some scientists won't even speculate.

The claim was a bit grandiose. In fact, the scientists had mapped only a certain portion of the epigenomes of two cell types (an embryonic stem cell and another basic cell called a fibroblast). There are at least 210 cell types in the human body, and each is likely to have a different epigenome.

Still, the potential is staggering. For decades we have stumbled around massive Darwinian roadblocks. DNA, we thought, was an ironclad code that we and our children and their children had to live by. Now we can imagine a world in which we can tinker with DNA and bend it to our will. It will take geneticists and ethicists many years to work out all the implications, but be assured: The age of epigenetics has arrived.

4
EPIGENETIC MARKS DEFINE CELL TYPES

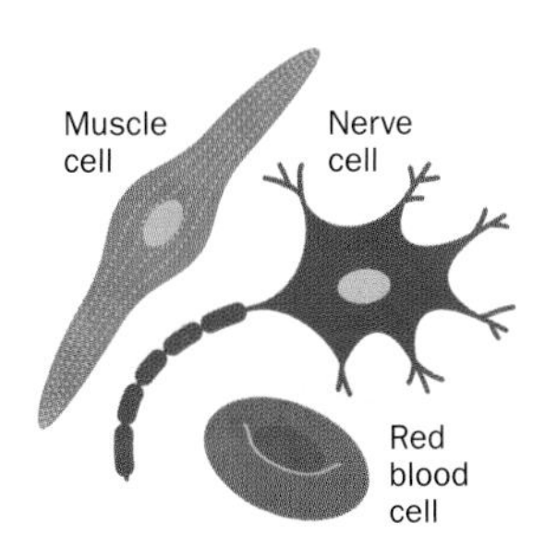

All cell types—muscle cells, nerve cells, and more—contain the exact same DNA. Epigenetic marks silence certain gene sequences and activate others so that nascent cells can differentiate. If the marks don't work properly, cancer or cell death is possible.

5
HOW ENVIRONMENT CHANGES THE MARKS

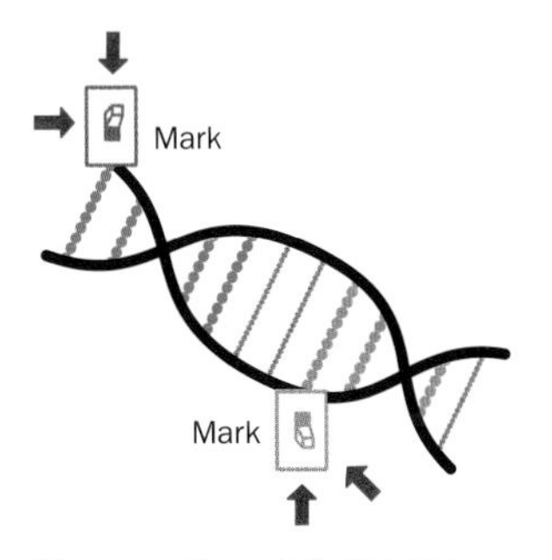

Stressors like a rich diet can activate epigenetic marks, modifying histones or adding methyl groups to DNA strands. These changes can turn genes on or off and may affect what gets passed down to your offspring.

6
WHAT EPIGENETIC CHANGES CAN MEAN

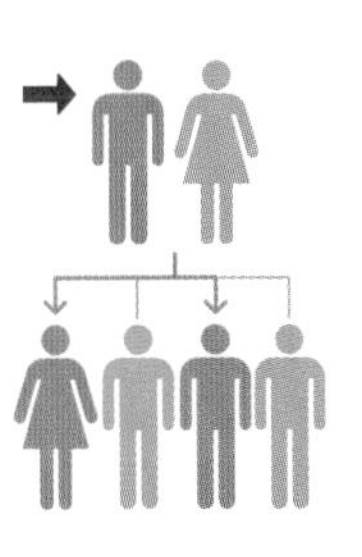

If you overstimulate genes for, say, obesity or a shortened life span, your kids can inherit these altered sequences. That could mean a lifetime of battling unfavorable gene expression.

THE GOOD NEWS
Scientists are learning to use epigenetic marks to switch off genes that lead to blood disorders and other diseases.

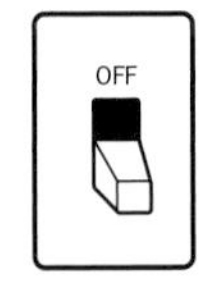

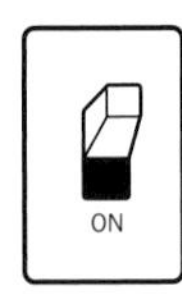

A Second Life for an Extinct Beast: Bringing Back the Aurochs

The only place to see an aurochs in nature these days? A cave painting. The enormous wild cattle that once roamed the European plains have been extinct since 1627. But this could soon change, thanks to the work of European preservationists and geneticists who are hoping they can make the great beast walk again. If they succeed, it will be the first time an animal has been brought back from extinction and released into the wild.

The aurochs was a massive creature, standing more than 6 feet tall at the shoulder and weighing more than a ton. But its size—and its prodigious horns—did not prevent it from being hunted to extinction. The hope for its resurrection now lies in its tame descendants, domesticated cattle. Scientists will first scour old aurochs bone and teeth fragments from museums to glean enough genetic material to recreate its DNA. Researchers will then compare the DNA with that of modern European cattle to determine which breeds still carry at least some of the creature's genes and create a selective-breeding program. If everything goes as planned, each passing generation will more closely resemble the ancient aurochs. The eventual goal is to replace the domesticated cattle that now graze in Holland's nature reserves with recreated wild cattle, restoring the countryside to a more natural state.

Other groups are trying to bring different animals back from extinction. In South Africa scientists are attempting to recreate the quagga, an extinct subspecies of the zebra, and in the U.S. breeders are trying to bring back a giant Galápagos tortoise that was killed off in the 1800s. In the world of creative genetics, never say never.

CLOSE KIN *A modern descendant of the extinct aurochs*

6.5 SHOULDER HEIGHT IN FEET OF THE AUROCHS. ON AVERAGE THE ANIMAL WEIGHED 2,200 POUNDS, MAKING IT AN ABUNDANT SOURCE OF MEAT—AND ACCOUNTING FOR ITS EXTINCTION.

Artificial Life Created in a Lab—Really

It's the ultimate science experiment: Take a handful of chemicals, mix them in just the right combination, and presto—life! After nearly 15 years of toiling in his labs in Rockville, Md., J. Craig Venter, co-mapper of the human genome, did just that, piecing together the entire genome of a bacterium and then inserting it into another bacterium. The cell booted up, and life was created. With Venter's breakthrough it's now possible to snap together genetic material to create a Legoland's worth of genetic combinations. Some of them could have robust industrial purposes, such as manufacturing bacteria to churn out valuable vaccine components or harnessing algae to convert carbon dioxide into biofuel. Those, of course, are the positive applications. Even Venter acknowledges that his technique could just as easily be used to generate dangerous mutants. His goal, he says, is to ensure that "the science proceeds in an ethical fashion."

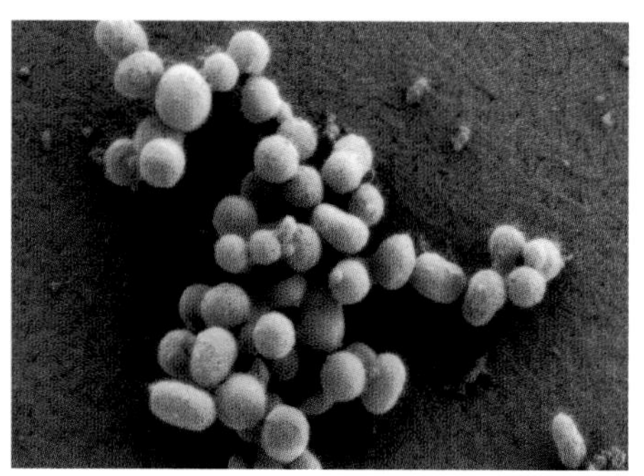

WRIT SMALL *Micrograph of bacterium*

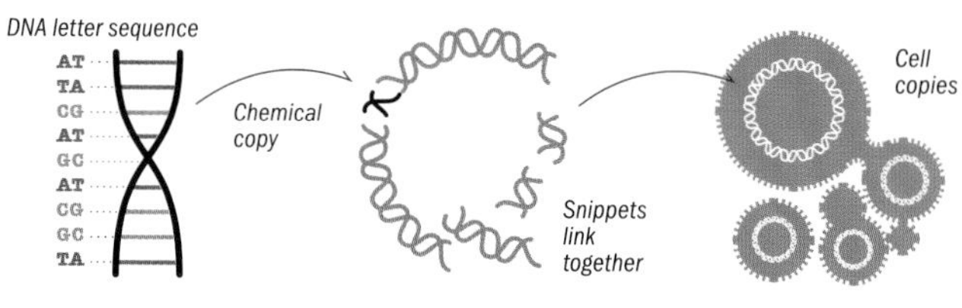

❶ SYNTHESIS A bacterium's four-letter-based genome is sequenced on a computer and duplicated using chemicals.

❷ CONSTRUCTION Snippets of DNA are fed into yeast cells that stitch them together until they form a complete, million-letter-long genome.

❸ ACTIVATION The synthetic DNA is substituted for that of a cell from another bacterial species, and the cell begins to divide.

FROM TOP: ALEX GRIMM/REUTERS/CORBIS; J. CRAIG VENTER INSTITUTE

ELI MEIR KAPLAN

Tracing Autism Back to Its Roots

An international group of researchers working in the Autism Genome Project Consortium has discovered a set of rare genetic variations that may increase susceptibility to autism spectrum disorders (ASDs). Past studies of twins and families have shown that though susceptibility to autism is primarily inherited, the complex genetic and environmental factors that actually lead to the condition have been difficult to untangle. In the new study, researchers conducted detailed genetic analyses of 996 people and found that those with autism were 20% more likely to have so-called copy-number variations—abnormalities in the number of copies—of specific genes. Most genes have exactly two copies, one inherited from the mother and the other from the father. More than two copies is the most fundamental kind of software problem. What's more, the anomalous genes spotted in the study were those involved in development and intellect, precisely where you'd expect to find irregularities leading to autism.

What has particularly energized the researchers is that all the genes linked to autism so far appear to affect similar biochemical pathways in the brain. That may offer shared targets for new drugs, so that even people whose autism has slightly different genetic or environmental causes may benefit from the same therapies. This understanding comes at a critical time, especially since experts are discovering that autism is a progressive disease, which in many cases can worsen if left untreated. Recent studies suggest that early intervention in infants—even before age 1—who are suspected of having ASDs could reverse or minimize some of the disorder's more severe symptoms. Still, genetic pathways are not the entire story. Even genes that are highly associated with autism don't always lead to the disorder; in most cases, these genes confer a vulnerability to the condition—but not a certain diagnosis.

Bacterial Genes May Determine Your Weight

Our intestines are teeming with trillions of bacteria that help us digest food. And a growing body of research suggests that having just the right population of these gut bugs may help keep the pounds off. Scientists are discovering that the gut flora of obese and normal-weight individuals are genetically different, and that their particular makeup is associated with whether calories taken in are turned into fat or fuel.

In the latest research, in mice, scientists led by Andrew Gewirtz at Emory University showed that normal-weight mice that were transplanted with the gut microbes from obese mice ended up getting heavier. The mice with the obese gut-bug profile also developed signs of metabolic syndrome, the constellation of symptoms including high cholesterol, hypertension, and diabetes that are associated with excess weight.

Previous mouse studies by Jeffrey Gordon at Washington University in St. Louis suggest that changing the diet can change the makeup of bugs. When one group of mice was fed a typical Western diet, high in fat and sugars, the mice tended to gain weight and grow more of a type of gut bacteria called Firmicutes and fewer of a type called Bacteroidetes. In mice given low-fat plant-based chow, the distribution of the two groups of bugs flipped and the animals remained lean. The shifts were dramatic and rapid, occurring in less than a day.

A GROWING BODY OF RESEARCH SUGGESTS THAT HAVING JUST THE RIGHT POPULATION OF BACTERIAL **GUT BUGS** MAY HELP YOU KEEP THE POUNDS OFF.

Taken together, the findings suggest that a "gut profile" could potentially serve as a diagnostic tool for identifying people who have a propensity for gaining excess weight. For starters, that could alert those folks that they need to watch their diet and get lots of exercise. The next step for scientists is to look for safe and reliable ways to change the gut microbe population in humans, leading to healthier flora, which in turn would mean healthier weight.

Who Discovered America? Not So Fast

Pity poor Leif Ericson. The Viking explorer may have been the first European to reach the Americas, but it is a certain Genoese sailor—Christopher something or other—who gets the glory. Part of the reason is that the record of Ericson's arrival here has been so sketchy. But a group of scientists may change that. Ten years ago Icelandic geneticist Agnar Helgason began investigating the origin of the island nation's population. Most of the people he tested carried genetic links to either Scandinavians or Britons. But a small group of Icelanders—roughly 350—carried a lineage known as C1, usually seen only in Asians and Native Americans. One of Helgason's graduate students studied the C1 line and was able to trace it to four women alive around 1700—far earlier than the time Asians began arriving in Iceland. That fact plus geographic proximity suggests that those women descended from a Native American. The most plausible way she got to Iceland would have been with a Nordic explorer like Ericson. If so, she arrived in Iceland around 1000, suggesting that Ericson met her in North America—and perhaps abducted her there—not long before. That beats Columbus's 1492 arrival by a cool 492 years. Ready for an Ericson Day parade next October?

LEIF ERICSON *An afterthought compared to Columbus, he may have gotten here first.*

FROM TOP: JOHN HERSEY; MACDUFF EVERTON/CORBIS

LIFE ON ICE *The $9 million Svalbard seed vault is home to the seeds for nearly half a million plant species. Wintertime temperatures average in the single digits.*

The Great Norwegian Seed Vault

The world's greatest greenhouse doesn't have a single plant. What it does have is seeds—more than 200 million of them, from roughly 490,000 species. They don't have much chance of growing, tucked away in a sort of hardened bunker on the far northern Norwegian island of Longyearbyen, where the Arctic cold helps keep them viable. But that's just the point. The Svalbard Global Seed Vault, as the facility is known, is the ultimate backup—a place to preserve the gene line of plant species from around the world in case climate change or natural disaster threatens to wipe a species out. The vault's remote location was chosen not just for the cold—though that's the key reason; seeds can remain dormant but alive for centuries if they're kept cool and dry. But should some human-engineered horror like nuclear war or a bioweapons attack occur, Svalbard could remain out of harm's way, capable of regenerating global agriculture when the dust clears.

For the moment, such a sudden disaster seems unlikely, but a slower one is already underway, in the form of the inexorable upward creep of global temperatures because of greenhouse gas emissions. To keep growing food as the world gets warmer, we'll need crops that are better equipped to withstand heat and drought. Breeders sifting through Svalbard's collection of seeds may discover tomorrow's food plants.

More Scars of the Holocaust: the Genes of Descendants

People who lived through the Holocaust were forever changed by what they experienced. In the 21st century many would be diagnosed with post-traumatic stress disorder (PTSD). As a result, a generation of children grew up in homes in which one and sometimes both parents were battling untold demons at the same time they were trying to raise happy kids. Studies of those second-generation survivors have found signs of PTSD in both their behavior and their blood—higher levels of the stress hormone cortisol, for example. Now research in animals suggests that a second generation's genes can be affected too.

The study involved raising mice from birth and continually separating them from their mothers until they were 14 days old. As adults, the animals exhibited PTSD-like symptoms such as isolation and jumpiness. More tellingly, their genes—most notably, one that helps regulate the stress hormone CRF and another that controls serotonin—functioned differently from those of other mice. When those mice had young, the pups exhibited the same anxious behavior and the same signature gene changes. What the study suggests about Holocaust survivors is likely true of people who survive horrors in Afghanistan, Iraq, or elsewhere, as well as those who grow up in abusive homes.

A SECOND GENERATION'S EMOTIONAL PROFILE IS NOT THE ONLY THING THAT CAN BE AFFECTED BY A PARENT'S TRAUMA; IT MAY ALSO BE **THEIR GENES.**

CLOCKWISE FROM TOP: PAUL NICKLEN/NATIONAL GEOGRAPHIC SOCIETY/CORBIS; JAN LUKAS/PHOTO RESEARCHERS; THE LIGHTHOUSE/VISUALS UNLIMITED

Hard-Wired for Fast Aging

Why do Brad Pitt and Angelina Jolie look the way they do and we, um, don't? One reason could be a little DNA sequence clustered near a human gene called TERC. The TERC gene is already known to produce an enzyme called telomerase, which helps regulate the length of telomeres—caps at the end of chromosomes similar to the plastic tips at the ends of shoelaces. Every time a cell divides, telomeres shorten, leading to a chromosomal fraying associated with aging and eventually the death of the cell. Shortened telomeres have been linked to a host of age-related illnesses such as heart disease and certain cancers as well. Now a British study published in the journal *Genetics* reports that people with one copy of the TERC gene had slightly shorter telomeres—short enough to be closer to those of people three or four years older. In other words, they were aging three or four years faster.

Scientists have long drawn a distinction between chronological and biological age—and doctors see that difference all the time in their practices, in 60-year-olds whose hearts and arteries are healthier than those of some 40-year-olds, for example. Lifestyle factors such as diet, exercise, and tobacco use are surely factors, but the TERC gene could be another. In a second study in *Nature*, researchers at Harvard Medical School found that they were able to switch on a telomerase gene in prematurely aged mice and reverse the aging process. The mice's organs regenerated, their shrunken brains increased in size, and their fertility was restored. In humans, manipulating the gene for telomerase could, in theory, slow aging too—or at least the development of age-related diseases. But there's reason for caution: Rapidly dividing, semi-immortal cells are also known as cancer cells, meaning that the search for eternal youth could yield an entirely different—and decidedly less pleasant—outcome.

LIVE LONG AND PROSPER *Families like the Hurlburts (below) are being studied by the National Institute on Aging for the secrets of their longevity.*

JASON GROW

NDM-1 IS A GENETIC MUTATION THAT RENDERS BACTERIA RESISTANT TO NEARLY ALL KNOWN ANTIBIOTICS.

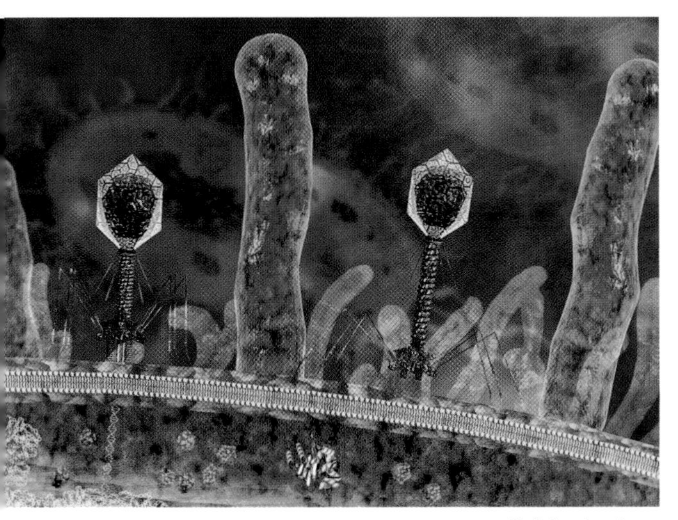

PORTRAIT OF A KILLER? *Bacteria just pulled ahead in the arms race with drugs, thanks to medical carelessness.*

Bacterial Supergene

Antibiotic-resistant bacteria aren't new, but the threat they pose is getting more urgent. The latest salvo is not actually a superbug but a genetic mutation called NDM-1. The initials stand for New Delhi metallo-beta-lactamase, since the variation was first detected in a patient who contracted an antibiotic-resistant infection in India. The gene variation has since turned up in Pakistan, Britain, the U.S., Canada, Japan, and Brazil. Bacteria with the NDM-1 variation produce an enzyme that neutralizes antibiotics, including last-resort medications known as carbapenems. Even scarier, say experts, is that NDM-1 can be passed like a secret survival code among different types of bacteria. So far, three species in the U.S. have picked up the genetic change, including *E. coli* and *K. pneumoniae.*

Most drug-resistant strains of disease emerge in hospitals, particularly in developing countries, where improper antibiotic use is high and the pressure on bugs to mutate to survive is great. Indeed, one of the reasons the supermutation first turned up on the Indian subcontinent is that antibiotic overuse is especially common there. Stronger antibiotics are not the answer, however, and only exacerbate the problem since bacteria will just find new ways to bypass them. The most effective way to fight resistance is to prevent it from occurring in the first place, by prescribing antibiotics only when necessary and ensuring that patients take them properly. When people do contract an NDM-1 infection, early diagnosis can make a big difference, as can careful use of the few antibiotic agents that do seem to have an affect on the infection.

IT JUST AIN'T SO...

Gene Tests Tell You Everything

Here's the good news: You almost certainly don't have the gene for cancer—or diabetes or Alzheimer's disease or any other illness we all dearly wish to avoid. Here's the bad news: That's because there's no such thing as "the gene for" any one of those ills. And that's a problem that's becoming more and more vexing as commercial gene tests allow consumers to map their genomes with a simple saliva swab and read their entire genetic blueprint in a single go.

Genes are a lot more complicated than they seem, and any physical or even emotional characteristic of a human is a result of a complex interplay of countless variables—often multiple genes, many external factors such as diet, and environmental toxins. In most cases a genetic test may reveal a few things—a susceptibility to an illness perhaps, or a somewhat increased likelihood of living to 100—but nothing about the certainty of those outcomes.

What worries the Food and Drug Administration is that consumers may be making medical decisions—everything from medication choices to preventive surgery—based just on a gene reading. The tests themselves have still not been proved safe and effective, not by the FDA's definition at least. And they surely have not been shown to provide anything like a precise medical prognosis.

What's more, errors are possible. One commercial tester recently conceded that it had mixed up the results of 96 customers, sending the wrong people the wrong profiles. The FDA—by nature a cautious organization—probably wouldn't mind if the mass market versions of the tests went away entirely, though that's not likely to happen. At the very least, consumers should think of their genome less as a reading of the future than as a recipe book of possibilities—and rely on their doctors to tell them what those possibilities mean.

DNA BLUEPRINT *A simple saliva swab can reveal a genome.*

Physics

SHAKING EARTH SHORTENS DAY ■ MAKING A BLOB OF LIGHT ■ THE INVISIBILITY CLOAK
SLOWING DOWN TIME ■ COMMUNICATING IN PHOTON CODE ■ TELEPORTATION ACHIEVED
NANOWIRES' VANISHING ACT ■ BEING IN TWO PLACES AT ONCE ■ LEVITATION IS REAL

The Machine That Could Explain Everything

The Large Hadron Collider is on a quest to answer the fundamental questions of the universe. It just may succeed—provided the scientists can keep it working.

By Michael D. Lemonick

Atoms are so ridiculously small that you could fit about 1,000,000,000,000,000,000 of them (that's a billion billion) on the head of a pin. Every last one of these minuscule objects is made mostly of empty space, with one electron or more buzzing around a nucleus—vastly tinier than the atom itself—made of protons and (usually) neutrons, which are in turn made up of even smaller particles known as quarks.

Given all of that tininess, it might seem odd that in order to study the inner workings of the subatomic world, physicists have built the most gigantic and complex scientific hardware on earth. But to the physicists themselves, it makes perfect sense. These impossibly small particles are bound together by forces far more powerful than gravity. You can't simply pry them apart with a toothpick.

Instead, you have to create knots of pure energy so intense that they mirror conditions a fraction of a second after the Big Bang, when the entire known universe was packed into the tiniest fraction of an inch and simmered at trillions of degrees. All the particles in the cosmos—everything that eventually formed into all the stars and galaxies we can see—condensed out of that miniature firestorm. To get a glimpse of the most elusive particles, you have to create a mini Big Bang so that they can condense out again.

For about a century now, physicists have been doing it by smashing atom fragments together at terrifically high speeds using particle accelerators (informally known, for obvious reasons, as "atom smashers"—a term physicists never actually use). And the biggest accelerator ever built is now humming away in an installation spanning the Swiss-French border near Geneva.

MAXIMILIEN BRICE/CERN

THINKING BIG *The appropriately named Atlas detector is just one of six such structures in the LHC. Its assembly of eight magnets weighs 7,000 tons and measures 84 feet.*

MAGNETIC MAW *The compact muon solenoid will explore physics at the Terascale—named for the one million million volts needed to do the work.*

CLOCKWISE FROM LEFT: MAXIMILIEN BRICE/CERN; CLAUDIO MARCELLONI/CERN; MAXIMILIEN BRICE/CERN; CERN; FABRICE COFFRINI/EPA/CORBIS

Called the Large Hadron Collider—LHC for short—the massive facility occupies an oval tunnel about 300 feet underground and 17 miles around, interrupted every couple of miles by sophisticated particle detectors that weigh thousands of tons each. The tunnel is lined with the world's most powerful magnets, whose task it is to send protons wheeling around the ring like tiny racecars traveling at nearly the speed of light—except that the protons are sent in clumps traveling in opposite directions, and they're meant to crash. It's a car race all right, but a kind of subatomic Indy 500 and demolition derby combined.

What comes out of these collisions of such unprecedented power, the physicists hope, will be species of particles theorists have been dreaming about for decades—some of which could solve mysteries most of us didn't even know existed. Perhaps the most prominent: Why does anything in the universe have mass? According to theorists, particles condensing out of the Big Bang shouldn't have had any (and if that sounds crazy, we already know of at least one massless particle: the photon, which is what light is made of).

The leading idea at the moment is that mass arises from particles known as Higgs bosons, which pervade all of space like a sort of undetectable cosmic molasses. What we think of as mass is actually the constant struggle of ordinary matter to slog through the stuff. The Higgs boson is so important that the Nobel Prize–winning physicist Leon Lederman named it the "God Particle."

Earlier generations of accelerators have tried in vain to create the Higgs. The Tevatron, outside Chicago, which Lederman ran for years, has recently shut down its own futile quest for the Higgs. The Superconducting Supercollider, which broke ground near Waxahachie, Texas, about two decades ago, might have found it but was canceled by Congress in 1993 before it could be completed. The LHC should find the Higgs—if it exists, that is, and if theorists are right about how powerful a collision it takes to create it. Nobody is absolutely certain on either of those points. If the Higgs doesn't exist, physicists will have to tear up years' worth of equations and start over.

The LHC could also yield the answer to another cosmically significant mystery. Starting in the 1970s, astronomers began to accept the idea that everything they could see in the universe—stars, galaxies, giant clouds of intergalactic gas—was only a fraction of the matter that was really out there. Something utterly invisible, and staggeringly massive, was pulling on the visible stuff, making it move around at speeds that it shouldn't on its own. To account for its gravitational effects on ordinary matter, there had to be many times more of this strange stuff, known as dark matter, than there was visible matter.

To this day nobody is sure what dark matter is. The leading theory suggests it's made of as-yet-unidentified particles gathered into huge clouds that surround and pervade every galaxy—including ours. Such ubiquity notwithstanding,

MEGA-MACHINE *Clockwise from top left: One part of the Atlas detector is lowered into place 300 feet below ground; a single stretch of the 17-mile oval tunnel in which packets of protons are accelerated at close to the speed of light on planned collisions; a computer simulation of one such crack-up—the different colors indicate different types of particles; inside the LHC control room.*

THE LARGE HADRON COLLIDER IS KIND OF A
SUBATOMIC INDY 500
AND DEMOLITION DERBY ROLLED INTO ONE.

dark matter has managed to escape detection. The LHC might help explain it. If it doesn't, astronomers and particle theorists will have to take two giant steps backward.

Third, and most weirdly, the LHC could provide the first evidence to back up string theory, a grand scheme that would meld relativity and quantum theory into a single unified set of equations explaining the nature and behavior of every form of matter and energy in the cosmos and all the forces they exert on one another. Most versions of string theory suggest that the universe has many more dimensions than the three we're used to—but the extra dimensions are all "compactified," so you can see their influence only on the tiniest scales. The LHC could, in principle, see particles disappearing into those extra dimensions.

For all the grandeur and ambition of the LHC, it is still just a machine—and like all machines, it has a tendency to break down. The engineers at CERN (the French acronym for European Center for Nuclear Research, the multinational organization that runs the accelerator) popped Champagne corks when the accelerator first went online, but they've spent an awful lot of time since simply trying to fix the darned thing. No sooner had the multibillion-dollar LHC started up in September 2008 than it had to be shut down because of problems with the electrical wiring in its powerful magnets.

The accelerator finally did go back to work again after a few months, but it's been limping along at half-power. CERN managers had planned to turn it off at the end of 2011 for a complete overhaul that would have taken more than a year. Now, however, they've decided to keep the machine going, even in its hobbled state. The reasoning, according to CERN research director Sergio Bertolucci in an official statement: "If Nature is kind to us and the lightest supersymmetric particle, or the Higgs boson, is within reach of the LHC's current energy, the data we expect to collect by the end of 2012 will put it within our grasp."

If Nature is less kind, it will clearly take longer. But if Nature is perverse, as so often happens when scientists push into completely unknown territory, the LHC could discover things that nobody has yet imagined. If that happens, Bertolucci may end up echoing the words of the physicist Isidor Isaac Rabi when a new, totally unexpected particle called the muon showed up in experiments: "Who," asked Rabi, "ordered that?"

CHILE'S STRUGGLE *The town of Iloca, 168 miles south of Santiago, after a monster quake in 2010 (top). A 15-story apartment building in Concepción lists drunkenly (above).*

Why a Shaking Earth Means a Shorter (or Longer) Day

If you wanted to make sure you got enough sleep the day after Chile's 2010 earthquake, you had to get to bed a bit earlier—about 1.26 millionths of a second earlier, to be precise. According to NASA, that's how much an Earth day should have been shortened by the subterranean upheaval that triggered the quake. Some basic physics explains why.

Every point on the planet takes the same 24 hours or so to complete a single rotation around Earth's north-south axis, but some points have to move faster than others to spin the full 360 degrees by the one-day deadline. That's because some parts of the planet are much bigger than others, at least in circumference. The Earth's equator is 24,901 miles around. The perimeter of the Arctic Circle, by contrast, is just 9,945 miles, and if you stand five feet from the North Pole, the circumference you inscribe as the Earth rotates is a scant 31.4 feet. Yet in all those places, it still takes 24 hours to complete a single rotation.

Earthquakes alter planetary speed in a two-step way. Shifting plates rearrange the distribution of the Earth's mass, causing it to bulge imperceptibly in spots it didn't bulge before and contract in others. That rearrangement should further shift the Earth's inclination, or figure axis (the axis around which the Earth's mass is balanced). In the case of the Chilean earthquake, the axis was shifted about three inches. That's not much, but the law of conservation of angular momentum requires that even under those subtly changed circumstances, motion must remain constant, which means that the planet must step on the gas (or the brake) to accommodate shifting mass. Post-Chile, that came out to the 1.26-millionth-of-a-second figure NASA calculated. The same rules apply to figure skaters, who change the speed of their spin by extending or tucking in their arms.

FROM TOP: MARTIN BERNETTI/AFP/GETTY IMAGES/NEWSCOM; JOE RAEDLE/GETTY IMAGES

A Blob of Light Means a New Kind of Matter

Say goodbye to one more simple truth of science—this time it's the one you learned in elementary school about all matter existing in a solid, liquid, or gaseous state. To those three, add one more: the blob.

Physicists had long suspected that a fourth state of matter existed—one they call a Bose-Einstein condensate, which is essentially extremely cold atoms of gas that cluster together in a bloblike state and behave like a single particle. As long ago as 1995, they created such a physical oddity, but the real challenge had always been to do the same thing with photons, or particles of light. Photons are particularly resistant to being supercooled, since they are massless, and if they lose too much thermal energy, poof! They simply disappear.

But researchers at the University of Bonn got around that by trapping photons between two mirrors separated by just one micron—a millionth of a meter—and mingling them with molecules of what amounted to a colored pigment, which absorbs and re-emits photons. Every time the photons were gobbled up and spat back out, they lost energy, and the mirrors kept them from bouncing away. Ultimately they cooled down to about room temperature—which is hardly the supercooled state the Bose-Einstein gas achieved but is still very cold, by photon standards at least. That was just enough to cause them to indeed clump into the condensate state.

A perfectly reasonable question that follows news of this achievement is: And? What good really is a Bose-Einstein photon blob? For one thing, it helps advance knowledge of quantum physics, revealing more about the behavior of all particles. It could also have eventual applications in the development of new, more flexible lasers.

The Man in the Invisible Cloak

When scientists speak seriously about something like a space-time cloak, they're either (a) extremely smart or (b) a tiny bit crazy. Professor Martin McCall of Imperial College in London is not crazy. The British physicist published a paper in the journal *Optics* describing the theoretical possibility of something he calls "metamaterials," fabric or other forms of matter that could be molecularly engineered to scramble the usual flow of electromagnetic energy. Light passing through them would emerge unevenly, creating gaps in time and space.

Okay, like so much about physics, that's hard to follow, but it has some real-world implications. All light slows down when it enters any material, but when the natural rhythm of that passage is disrupted, it means that some light would arrive at a destination—say, your eyes—earlier than it would otherwise, and the rest might arrive later. McCall, only half-jokingly, imagines a criminal wrapped in a cloak of metamaterials entering a room, robbing a safe, and leaving while a surveillance camera reveals nothing amiss. One tiny glitch in the plan: Given the speed at which light travels, invisibility for even a few minutes would require a cloak about 320 million feet long.

A less fanciful and decidedly more practical application would be in data processing. Metamaterials could speed and slow the transmission of information, preventing data bottlenecks and even improving encryption.

A SPACE-TIME CLOAK

COULD CREATE A SORT OF CORRIDOR THROUGH WHICH LIGHT WOULD MOVE AT DIFFERENT RATES.

HIGHER MEANS FASTER

GRAVITATIONAL TIME DILATION IS THE PRINCIPAL RULE UNDER WHICH ALTITUDE MAKES CLOCKS SPEED UP.

Speeding Up to Slow Down Time

Einstein didn't live to see *Planet of the Apes*, but he did articulate its central premise—that if you travel fast enough, time moves slower for you than it does for people standing still. Few physicists have ever doubted that idea, but now it's been demonstrated.

In an elegant experiment at the National Institute of Standards and Technology (NIST) in Boulder, scientists developed two atomic clocks, each built around a single aluminum ion trapped in an electric field. How this works to mark time is complicated, but there's little mistaking how well the system does its job: The clocks are accurate to within one second over 3.7 billion years.

With both clocks running, the scientists ran a series of experiments, keeping one stationary while the other was transported about at speeds as low as 22 mph. In all those cases, time on the moving clock moved slower than it did on the unmoving one. Time also moved at different rates when clock one was hoisted as little as 13 inches above the other, proving another Einsteinian theorem about gravity fields and speed. But don't count on 13-inch platform heels to keep you young. If you live to be 80, they would add only 90-billionths of a second to your life.

Need to Send a Coded Message? Write With Photons

China's military keeps getting more sophisticated, and recently it achieved one of its most impressive breakthroughs—involving a technology that has nothing to do with weapons. Military physicists, writing in the journal *Nature Photonics*, announced they had composed coded, Morse code–like messages with photons—the essential particles of light—and transmitted them at light speed via a blue laser.

Why is this fantastically complex method superior to e-mail or radio? Because, says Mathew Luce, a researcher at the Defense Group Inc.'s Center for Intelligence Research and Analysis, it "cannot be cracked or intercepted. [It's] secure communications guaranteed by the laws of physics." What makes the code so ironclad is something called the Heisenberg uncertainty principle, which states that the mere act of measuring a particle alters it. So if

the photons in the laser beam are observed by a third party, the particles themselves will be rearranged. When that happens, the sender and receiver would be immediately informed that someone was snooping.

Luce says that a blue laser—instead of an infrared one like the U.S. has been testing for a similar technology—was chosen with China's growing submarine fleet in mind, since blue lasers penetrate farther underwater. Soon Chinese satellites could be able to communicate with submarines without their needing to surface or give away their location by breaking radio silence. In the military showdowns of the future, it seems, brainpower may become more important than firepower.

CLOCKWISE FROM RIGHT: PASIEKA/SCIENCE PHOTO LIBRARY/CORBIS; CORBIS (2)

Beaming Up Achieved for Electrons. Is Scotty Next?

Inching our reality ever closer to *Star Trek*'s, scientists at the University of Maryland's Joint Quantum Institute successfully teleported data from one atom to another. A landmark in the brain-bending field known as quantum information processing, the experiment doesn't have the cool factor of body transportation; one atom merely transforms the other so it acts like the first. Still, atom-to-atom teleportation has implications for creating supersecure, ultra-fast computers.

The atomic teleportation concept is known as "quantum entanglement," and it's the same one the Chinese are using with their blue laser messaging. The idea is that electrons vibrating in synchrony can be separated by anything from a few inches to a billion light-years and still be energetically connected. Jiggle one and the other responds—with the information transferred at or faster than the speed of light. That idea had been only a theory at anything beyond atomic scales until the Maryland team achieved it between two electrons separated by about three feet. In the world of physics, the teleportation of information is the same as the teleportation of matter, so in a crude way, what happened in the lab is the same as what happens on the bridge of the *Enterprise*. But there is a catch: At a human scale, Mr. Spock wouldn't be beamed up from the surface. He'd be atomically annihilated on the planet and then be rebuilt on the ship. That's got to hurt.

Another Invisibility Sighting

Berkeley did nothing to change its rep as one of America's flakier places when scientists on the local campus of the University of California announced that like the folks at the Imperial College of London, they'd invented an invisibility cloak. But it was hard physics and complex optics at work, not something illegal or brain-altering. Using nanowires grown inside a porous aluminum tube to create a sheeting 10 times thinner than a piece of paper, they proved that they could wrap an object in the material and bend light waves around it, making it effectively invisible. All of the usual caveats apply: The process is experimental, the cloaking is fantastically fragile, the costs would be prohibitive for anything practical. Still, we now live in a world in which invisibility is a possibility. That's a good thing, right?

FROM TOP: PARAMOUNT/EVERETT COLLECTION; EVERETT COLLECTON

BAOBA IMAGES/GETTY IMAGES

.18

THE FRACTION OF A DEGREE BY WHICH A TEAM OF PHYSICISTS FELL SHORT OF COOLING AN OBJECT TO THE –459.67 DEGREES FAHRENHEIT TEMPERATURE KNOWN AS ABSOLUTE ZERO

An Object That's Here—And There Too

Be glad you're not Erwin Schrodinger's cat. The 20th-century Austrian physicist devised a paradox in which a cat was placed in a box with a vial of poison that would be shattered if an electron was in one position and remain intact if it was in another. The paradox comes from the fact that electrons—thanks to the concept of superposition—can be in two places at once, so Shrodinger's cat would be both dead and alive. No one has ever run the experiment, but the concept of superposition has always been accepted—for electrons. Now physicists at the University of California at Santa Clara have pulled off the same feat with an object big enough to be seen—just barely—with the naked eye.

The researchers fabricated a tiny, resonating aluminum disk made of only about a trillion atoms—which is really, really small. They cooled it to about 0.1 degree Kelvin, which is a brisk –459.49 degrees Fahrenheit. That's awfully cold, but still just shy of the –459.67° F point known as absolute zero, at which there's effectively no more heat for an object to lose.

The purpose of the superchilling was to eliminate stray activity and then measure if the disk could be both resonating and not resonating in the same moment. A superconducting quantum bit—or qubit, which is a sort of quantum thermometer—then took the temperature of the disk and found that it had indeed achieved a state of superposition. Confused? If so, you're part of a very big club. Quantum physics is headache-inducingly complex, but the possible applications of it are easier to grasp. Mastering superposition technology could make quantum computing—in which electron states take the place of charged states on a microchip—possible. Think your iPod is small now? Just wait.

Levitation Achieved—Really

Physics always seems to want to come out and play. Just when this most technical of sciences starts to become impossibly arcane, it goes goofy on you. That happened in a big way when a team of researchers announced in the journal *Nature* that they'd mastered a new way to levitate physical objects.

Everything in the universe—metals, gases, dogs, doughnuts—is made of materials with positive and negative charges. Opposite charges attract each other; identical charges repel each other. What prevents us from sticking to anything with an opposite charge is that all these forces have to be properly aligned before you can see them at work. In the *Nature* experiment, the team placed a microscopically small sphere of gold on a glass surface. Gold and glass get along well enough and under the right circumstances will attract. But what they both like a whole lot more is a liquid called bromobenzene. When the researchers introduced a little bromobenzene to the other two materials, they both began drawing so much of it that the gold began to rise above the glass. In effect, it levitated on a thin bromobenzene film.

Increasingly, nanoengineers are working to develop medical devices, electrical switches, and more made up of microscopic parts that float above one another on thin films of other materials. This increases efficiency, reduces friction, and allows the hardware to be built to finer tolerances and at tinier sizes. Even at its most fanciful, physics, it seems, can play around for only so long before it gets back to serious work.

IT JUST AIN'T SO ...

HARMLESS AND LOVELY *The LHC, lit by floodlights at night*

Bulletin: Machine Didn't Eat Earth

Good news: The Large Hadron Collider (LHC)—the particle accelerator straddling the Swiss-French border —didn't destroy the world. Before the LHC went into operation in 2008, critics warned that it might create a mini black hole that could expand to planet-eating proportions. On Aug. 26, 2008, Otto Rossler, a German chemist, filed a lawsuit in the European Court of Human Rights arguing—apparently without irony—that such a scenario would violate the right to life of European citizens and pose a threat to the rule of law. Five months earlier, two American environmentalists had filed a lawsuit in Federal District Court in Honolulu seeking to force the U.S. government to withdraw its participation from the experiment. The lawsuits in turn spawned several websites, chat rooms, and petitions—and they led to headlines about the impending end of the world.

Scientists did what they could to allay the fears. Researchers at the European Organization for Nuclear Research, which runs the LHC, published a safety report, reviewed by a group of external scientists, ruling out the possibility of dangerous black holes. It said that even if tiny black holes were to form—a big if—they would evaporate almost instantaneously because of Hawking radiation, a phenomenon named for the British physicist Stephen Hawking, whose theories show that black holes not only swallow up light, energy, and matter but leak it all back out at an accelerating pace.

Fear of the scientific unknown is not new. When the locomotive was conceived, even some engineers predicted catastrophe resulting from the human body's inability to withstand the strains of high-speed travel. The word *vaccine* comes from the Latin word for cow, *vacca* —and the first vaccinations, against smallpox, used bovine ingredients. That sparked fear that the injections would turn humans into cows. No cow-people resulted from the shots—and the LHC, similarly, didn't destroy the world. Fear is natural—but reason can be powerful.

FROM LEFT: GRANGER COLLECTION; MARTIAL TREZZINI/EPA/CORBIS

Medicine

THE ARTIFICIAL OVARY ■ POWERFUL PLACEBOS ■ BANKING CORD BLOOD
A VACCINE AGAINST COCAINE ■ THE POWER OF SOCIAL NETWORKS ■ A PILL TO PREVENT HIV
EPIGENETICS AND OBESITY ■ ANTICANCER VACCINE ■ A HEART-ATTACK BLOOD TEST

Stem Cell Science Races On

The days of ethical warfare over stem cells seems so 20th century. New advances make the science better, faster, and less politically fraught.

By Alice Park

Twenty-five days. That's how long it took Dr. Shinya Yamanaka of Kyoto University to undo more than three decades of the exquisitely programmed biology packed into a middle-aged woman's cheek cell—and just maybe change the world. In those three weeks Yamanaka turned back the clock on an aging cell. In the ultimate feat of reprogramming, he tricked it into acting like that wonder of cellular shape-shifting, the embryonic stem cell—capable of dividing, developing, and maturing into any of the body's more than 200 different cell types and helping treat many of its myriad diseases.

What's remarkable is that Yamanaka wasn't the only one. On the same day in 2007 that he published his milestone in the journal *Cell*, James Thomson, a pioneering molecular biologist at the University of Wisconsin who in 1998 discovered the first human embryonic stem cells, reported similar success in *Science*, reversing development in foreskin cells from newborns. What made Yamanaka's and Thomson's work so remarkable is that it entirely sidestepped the use of living embryos. That raised the tantalizing possibility that the long-raging stem cell debate was at last over. Cells generated by Yamanaka's or Thomson's method would be ideal, and not just because they were free of political and moral baggage: Since they originated in a patient, they could also give rise to any type of body tissue and then be transplanted back into the donor with little risk of rejection. In the four years since their landmark achievements, the promise has only grown greater—mostly because the science has advanced even further.

It wasn't always obvious that such direct reprogramming of adult cells into "induced pluripotent stem cells," or iPS cells, could ever be achieved. There were perhaps

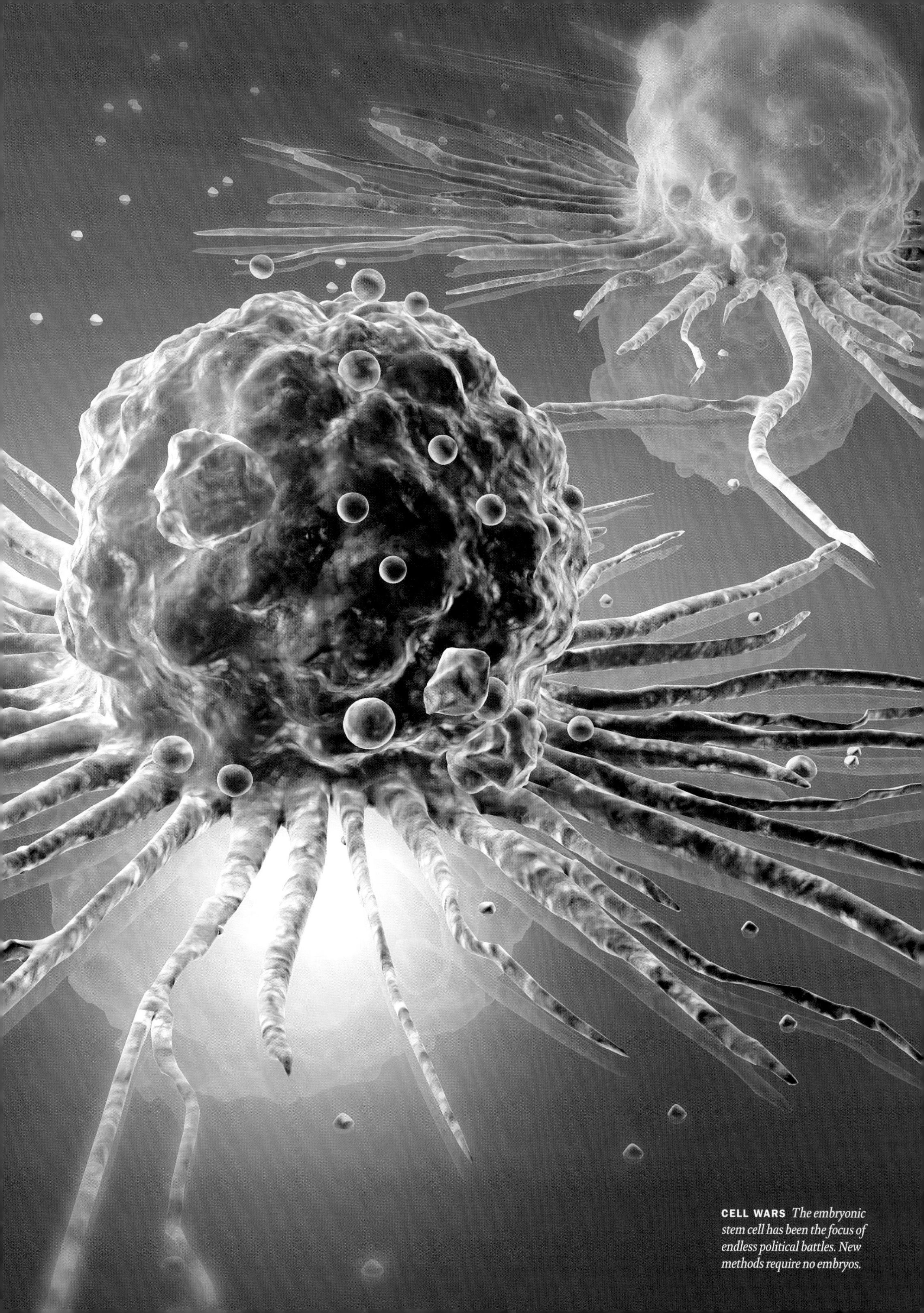

CELL WARS *The embryonic stem cell has been the focus of endless political battles. New methods require no embryos.*

100 protein factors and an unknown amount of gene jiggering that would have to be done to recalibrate the cell properly, and all of it would have to be discovered by trial and error. "When I started this work," says Thomson, "I thought it would be a 20-year problem."

Fortunately, it was easier than that. The fountain-of-youth factors that he and Yamanaka used turned out to be well-known genes active in early development. Both scientists relied on inserting a separate set of four genes into aging cells, using the most efficient genetic bullies around: viruses, which penetrate a cell's membrane and insert new genetic software into its nucleus. The technique is comparatively efficient—about one stem cell line per 5,000 cells in Yamanaka's case, or one stem cell line for each cultured petri dish of cells. While that may not sound impressive, in the hit-or-miss world of stem cell research, it virtually counts as a sure thing. "The efficiency puts it well within the realm of technical feasibility," Dr. George Daley, a stem cell expert at Children's Hospital in Boston and the Harvard Stem Cell Institute, said at the time.

Still, that didn't mean the cells were ready for transplant into patients. The viruses used to ferry the genes were retroviruses and lentiviruses (the families that include HIV), which can introduce genetic mutations that cause cancer. But as Dr. Douglas Melton, co-director of the Harvard Stem Cell Institute, speculated when the Yamanaka and Thomson papers were published, "Eventually we may not need to add genes or viruses at all to cells. It will be possible to find chemicals that tickle the cells to turn the right pathways on."

That eventually occurred in 2010. In the same week that an appellate court allowed the government to continue its funding of human embryonic stem cell studies after a judge had halted the grants, researchers at Children's Hospital in Washington, D.C., and the Harvard Stem Cell Institute made news of their own, announcing that they had indeed developed a nonembryonic stem cell that required the use of no viruses or added genes.

The team, headed by Harvard pathologist Derrick Rossi, began with the simple truth that genes are nothing more than stretches of DNA that a cell "reads" by turning the DNA into messenger RNA (mRNA), a differently coded version of the gene that can then be made into proteins such as hormones or enzymes or other compounds the cell needs to survive. Instead of introducing genes into the cell, as Yamanaka had, Rossi thus decided to skip straight to the RNA step. If he could provide the cells with enough snippets of the right RNA corresponding to the Yamanaka genes, then the cell would still take the RNA and make it into the four gene-produced factors that Yamanaka had found were essential for reprogramming the cell.

Nice idea, but Rossi's team immediately encountered a problem. Cells have a built-in defense against foreign RNA (viruses, after all, are made up of RNA), so when the scientists first tried the experiment, the cell simply chewed up the added RNA. Some cells even died in the effort.

ILLUSTRATION BY ROBERT A. DI IESO JR.

The Science of Stem Cells

Stem cells obtained from embryos

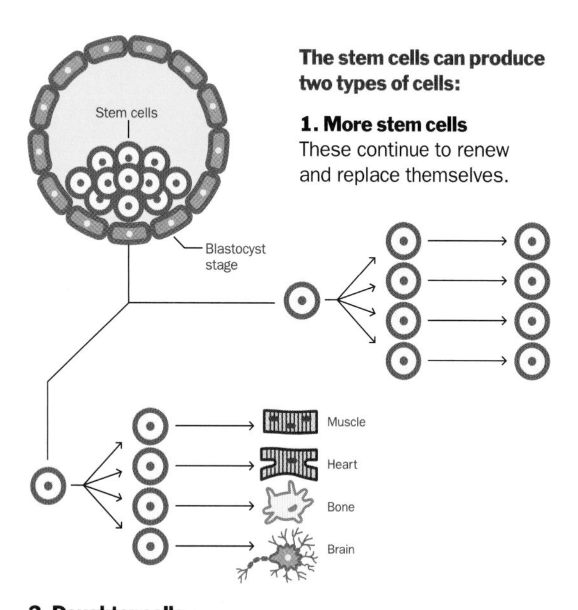

2. Daughter cells
These divide to become any specialized cell in the body.

Stem cells obtained from adult cells

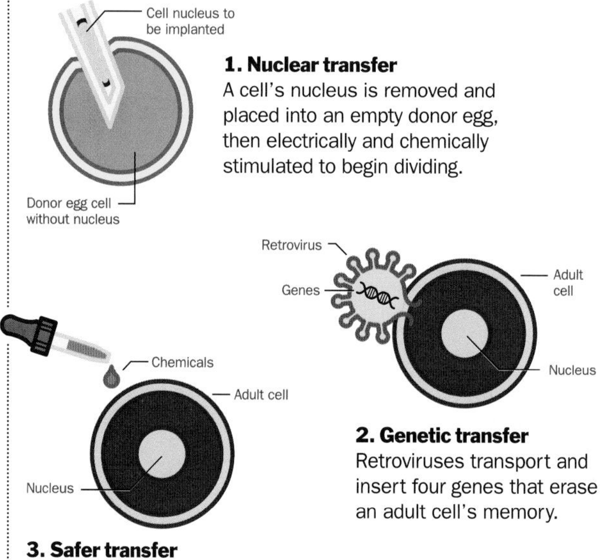

1. Nuclear transfer
A cell's nucleus is removed and placed into an empty donor egg, then electrically and chemically stimulated to begin dividing.

2. Genetic transfer
Retroviruses transport and insert four genes that erase an adult cell's memory.

3. Safer transfer
Introducing genes can cause tumors, so scientists are replacing them with chemicals or safer viruses.

The group persisted and found a way to attach a stretch of additional code onto the RNA pieces that would make them invisible to the cell's destructive forces. Once they did that, the cells happily pumped out enough of the factors to reprogram the cells and turn them back into an embryonic-like state. In fact, the system worked better than Rossi had expected. The mRNAs that he made and introduced into the cells resulted in up to a 100-fold increase in efficiency in generating iPS cells.

"We noted that we weren't just getting iPS cells, but getting very, very high efficiency," Rossi told reporters during a teleconference. "This was unexpected."

The results impressed Melton too—enough to persuade his colleagues overseeing the institute's iPS facility to switch over to the new technique for making cell lines. "This solves to a large extent the problem of efficiency," he said during the teleconference. "The technique makes it much faster and much more user-friendly. So we are going to turn over our entire iPS core to this method to efficiently make stem cells from patients with all sorts of different diseases so that we can begin drug screening and studying the causes of these diseases."

In a written response, Yamanaka also expressed optimism about the technique's potential for generating iPS cells that are safer for human patients. "I think that the method to generate iPSCs, described in the paper, seems promising in inducing clinical-grade iPSCs and would like to have someone in my lab try the protocol."

The new strategy could become the basis for a standard way of generating colonies of cells that are turned into healthy versions of diseased tissue that can then be transplanted into patients, says Melton. So far, the new RNA-based iPS cells appear to be more similar to human embryonic stem cells than they are to stem cells made the Yamanaka way. Rossi believes that may be due to the fact that the Yamanaka cells contain viruses, which may affect their ability to reprogram themselves. But it will take additional studies to confirm that the new iPS cells work the same way as human embryonic stem cells, which are currently the gold standard.

Even if the cells do prove to be safe and viable, Rossi admits that his approach doesn't mean that the Yamanaka method will become obsolete. For one, Rossi's technique requires the daily addition, including over the weekend, of a dose of RNA to the cells over a period of two to three weeks, while Yamanaka's viral method requires a one-time introduction of the virus-containing genes. For certain lab-based studies that won't involve human patients, he says, the Yamanaka method may be more convenient.

It's also not clear how long-lived Rossi's iPS cell lines are, since they have been around only long enough to divide a few dozen times. Yamanaka noted that iPS cell lines vary slightly in their ability to turn into different types of cells, and that more studies need to be done before a standard method for making the cells is established. Caveats aside, the science roars on—and this time free of politics.

Once generated from patients, stem cells can do two tasks.

1. Treat a disease directly
New genes can fix defects in the cells grown from stem cells. The repaired cells are transplanted into the patient, where they find and replace damaged tissues.

2. Study a disease in progress
Stem cells from a patient will develop the disease in question and show where sick cells go awry.

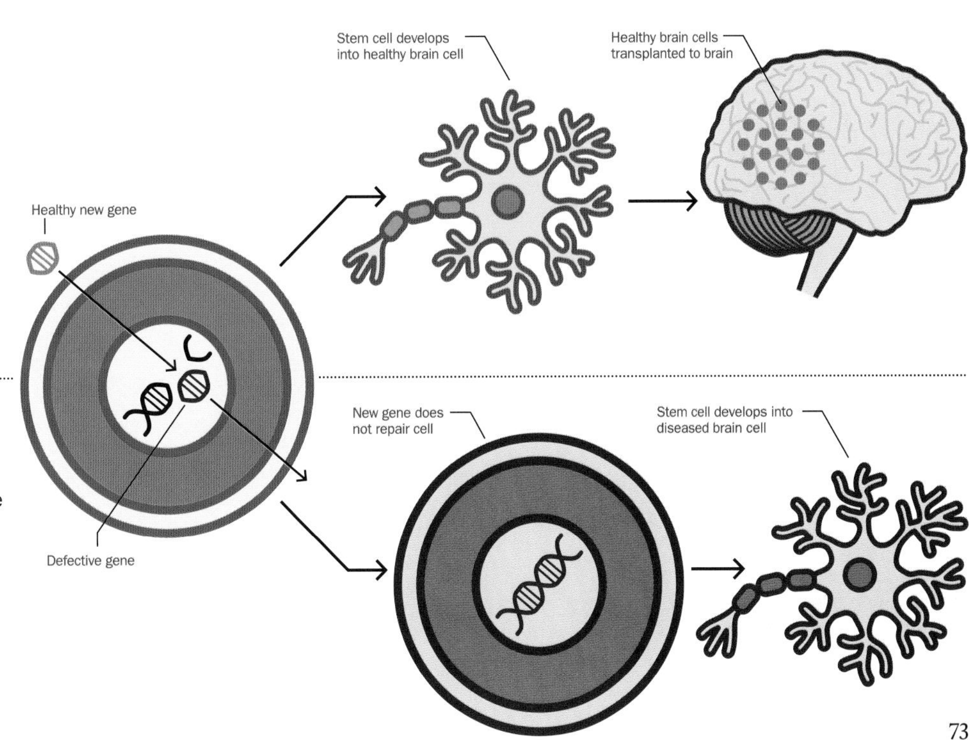

The Growing Power of the Fake Pill

Physicians have long believed that some form of deception is essential to the placebo effect. After all, if you tell people that you're giving them a fake drug, why would they respond by getting better? But a new study conducted at Harvard University suggests that it may one day be possible to use placebos in everyday medicine without misleading patients into thinking they might get active treatment. Patients suffering from irritable bowel syndrome (IBS) were divided into two groups. The members of one group received no treatment; the others got a placebo—and that fact was disclosed to them. But they were also told that the pills "have been shown in clinical studies to produce significant improvement ... through mind-body self-healing processes." The result: 35% of the no-treatment group experienced an improvement in symptoms, compared with 59% in the placebo group.

What could account for a sugar pill working such magic even when a patient knew it was fake? Part of the reason might be that, phony pill or not, subjects were told it could help them, which means that some deception was still involved. But the statistical gap between the two groups was so huge that there is likely something more at work. "This is just the first step to see whether there are ethical ways to harness the placebo effect," says Harvard's Ted Kaptchuk, a co-author of the study. "In order to be clinically applicable, it would have to be replicated in a much larger sample and continued for a much longer period of time." That work, surely, will be forthcoming.

The Newest Way to Improve IVF Success Rates: an Artificial Ovary

In more good news for those struggling with infertility, scientists reported success in creating an artificial ovary that could one day nurture immature human eggs outside the body. Researchers led by a team at Brown University managed to coax three primary ovary cells donated by patients to become a 3-D structure resembling an ovary. In the lab the cell types interacted with one another and functioned for all practical purposes like a real ovary, even successfully maturing a human egg from its earliest stages in the follicle to a fully developed form.

Most immediately, the structure could help IVF technicians improve their success rates. Currently when women donate eggs for a cycle of IVF, they provide a range of both mature and immature eggs; the less developed ones are less likely to be fertilized and go on to become viable embryos. But by allowing technicians to mature those eggs in the lab, the artificial embryo might help each IVF cycle become more efficient, boosting the odds that it will lead to a pregnancy and eventual live birth. In addition, the artificial ovary could help women with ovarian disease, who are unable to produce mature eggs, take advantage of IVF to have biological children of their own.

30% TO 35%

THE RATE OF LIVE BIRTH FOR EACH IVF CYCLE AMONG WOMEN UNDER THE AGE OF 35. AMONG WOMEN OVER 40, IT'S JUST 6% TO 10%.

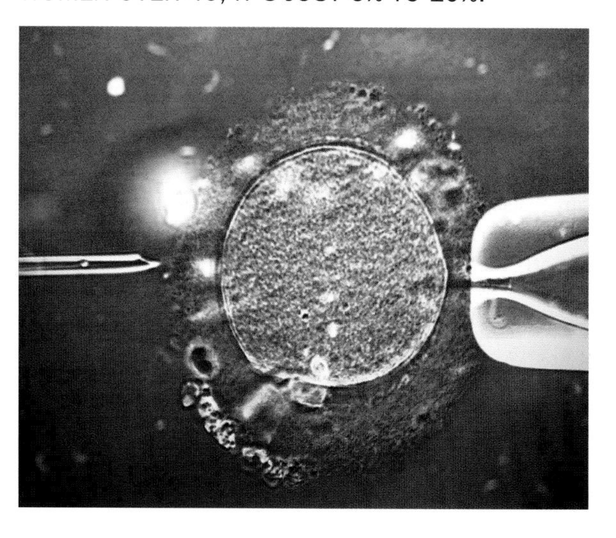

FROM LEFT: IMAGE SOURCE/CORBIS; GILLES PODEVINS/SCIENCE PHOTO LIBRARY/CORBIS; MAX AGUILERA-HELLWEG

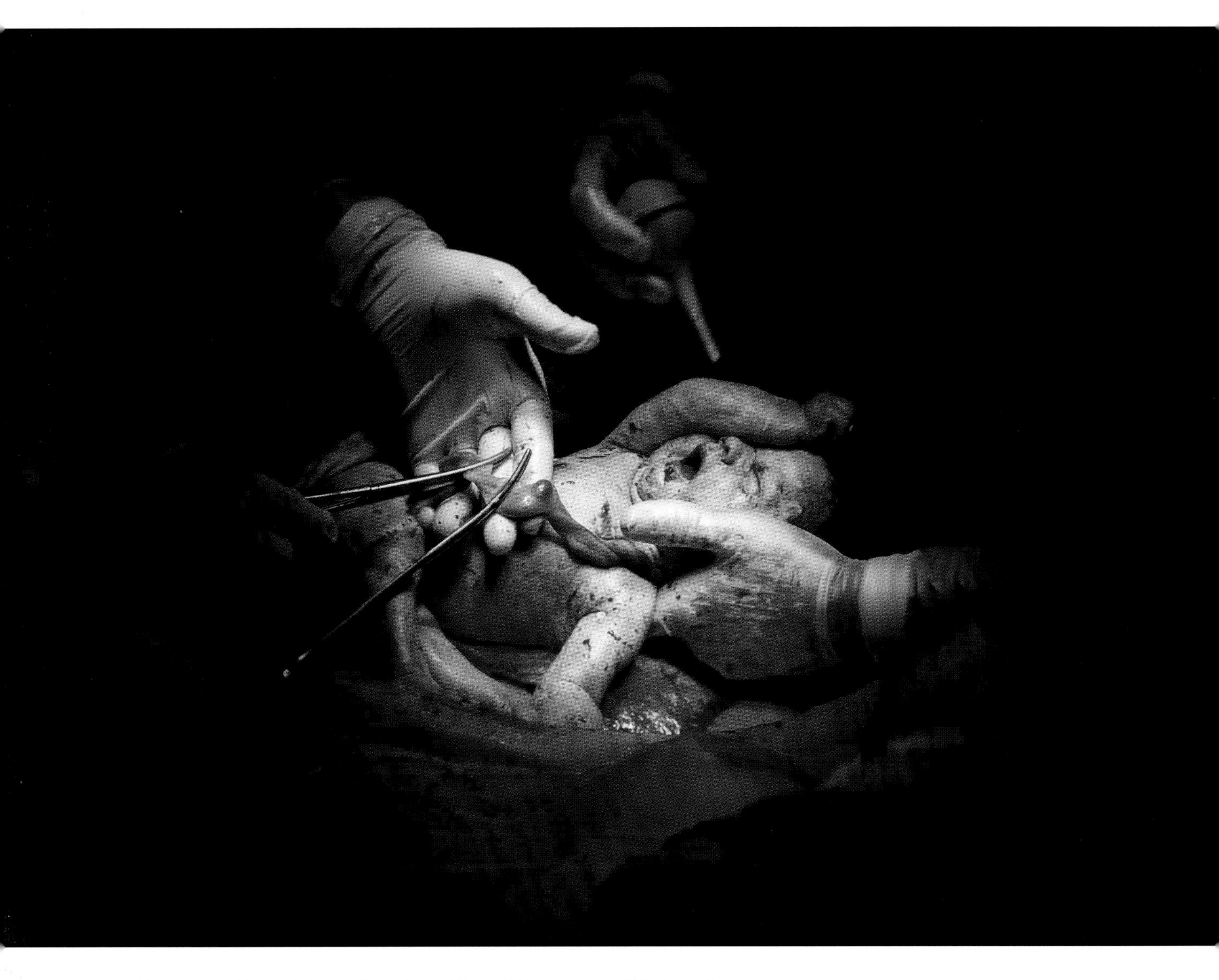

Saving More Lives by Banking More Cord Blood

As their due date creeps closer, many pregnant women pack a go-bag for the hospital: toothbrush, iPod, cute baby outfit. But recently, savvy mothers-to-be have started tucking in one more important item: a kit to collect and donate the blood in their baby's umbilical cord. Cord blood is a source of stem cells, already being used in therapy regimens for patients with cancer, sickle-cell anemia, immuno-deficiency, marrow failure, and genetic diseases that call for transplants, yet experts estimate that 99% of this potentially lifesaving resource gets thrown away postpartum.

Private cord-blood banks aggressively advertise their services to pregnant women and charge a bundle—some $2,000 for initial processing and about $125 a year for storage after that. The blood can be used only by the donor's family. Public banks—which make their contents available to anyone who is a close enough match—are an improvement, but there are only 19 of them in the U.S., and until recently the only way women could donate to them was to give birth in one of the 175 or so affiliated hospitals that have a system in place to collect and transfer cord blood.

That's where the Public Kit Donation project comes in. Three hospitals are piloting a federal program that allows women to mail in cord blood from anywhere in the continental U.S. for inclusion in Be the Match, the national cord-blood registry. It costs nothing to donate. The tab for collection, processing, and storage can run to $2,500 but is being split by the government and the participating banks. Whenever matches are made, the banks are paid by recipients or their insurance companies, the same as they would be for a donated pint of blood from your arm. Cord blood costs far more than ordinary blood—up to $35,000 a unit. But how do you put a price on a life saved? Easy: You don't.

94%

Share of patients who said it was perfectly all right for doctors to ask them about their religious beliefs. Plenty of doctors agree that if health-care providers suggest complementary care like acupuncture to some patients, why not faith and prayer to others?

How Moods, Disorders, and Habits Become Contagious

You can catch a cold; you can catch the measles. But what about depression? Can you catch a case of that? What about obesity or heavy drinking or even loneliness? Yes, yes, yes, and yes. Increasingly, epidemiologists and other scientists are discovering that social networks can spread bad and good behaviors and moods like any other contagion. More remarkably, you don't have to catch a case of, say, the blues from someone you know directly. It can be enough if your friend has a friend who's depressed.

The leader in the field of this new kind of social networking is Nicholas Christakis, a physician and sociologist at Harvard University. Christakis's most headline-making study was one in which he tracked 5,000 people over 20 years and found that when people who are close to us, both in terms of social ties (friends or relatives) and physical proximity, become happier, we do too—and that extends out to a third degree of separation as well. Christakis has also done the most noteworthy work showing that smoking and obesity can be just as socially infectious.

A later study, also at Harvard, made the distressing finding that eating disorders spread among teens in the same way. Girls in Fiji who had access to television were at increased risk of developing unhealthy eating habits—no surprise, given the idealized bodies seen on TV. But their friends who never watched TV were also at elevated danger. All this work may lead to new ways to stop unhealthy contagions and foster good ones.

A Possible Vaccine—Against Cocaine

Addiction is undeniably a disease—so as with other diseases, could you be vaccinated against it? The answer, according to a new study, is yes—at least if the addictive substance in question is cocaine. Researchers attached a cocaine-like molecule to an ordinary cold virus and injected it into laboratory mice. The animals' blood produced antibodies to the altered virus. When the blood was then mixed with actual cocaine in a test tube, those antibodies bound to the drug. This suggested that in the mice's systems, the antibodies would do the same, tying up the cocaine before it could reach the brain.

To test that idea, the researchers gave both immunized mice and control mice high doses of actual cocaine. As they anticipated, the immunized mice showed much less hyperactivity and other signs of intoxication than the ordinary mice did—and the effects of the vaccine lasted 13 weeks on average. Even if the vaccine proves itself in humans, experts on addiction warn that it's not a panacea. Addicts could try to overcome the immunological blockade by increasing their drug intake, leading to dangerous overdoses.

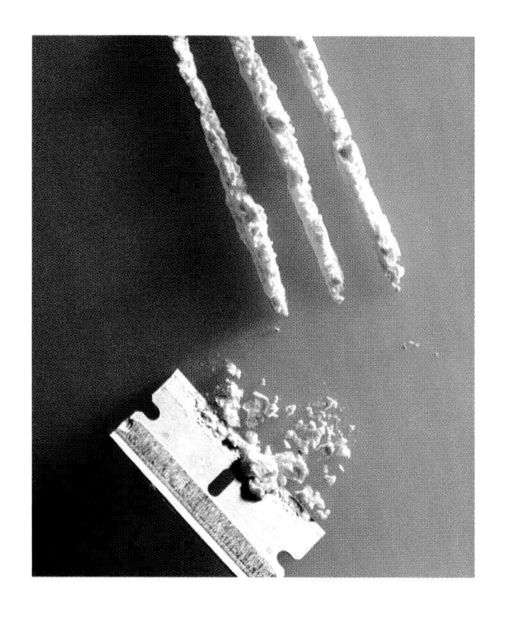

CHRISTIAN NORTHEAST; LESTER LEFKOWITZ/CORBIS

GIDEON MENDEL/CORBIS (3)

A Pill Can Fight HIV Before You Even Contract It

Antiretroviral drugs have turned the AIDS epidemic around by thwarting the virus in HIV-positive patients. But new research suggests that this powerful treatment may have another benefit—as a shield against infection in healthy people.

In a trial involving nearly 2,500 HIV-negative but high-risk gay men in six countries, researchers found that a combination antiretroviral pill called Truvada reduced the risk of HIV infection by 44% compared with a placebo. When the scientists looked more carefully at the study volunteers who took the medication most faithfully on a daily basis, they found that the risk of contracting HIV was even lower—73% lower than that of the placebo group.

More studies will need to confirm the benefit of antiretrovirals in the prevention of HIV, and public-health experts warn that even if the results hold up, it would not replace the best method of prophylaxis: safe sex and consistent use of condoms. That's because the way so-called pre-exposure prophylaxis, or PrEP, works is to load up high-risk people with HIV-disabling antiretroviral drugs before exposure to the virus, which allows the medication to hit HIV as early as possible. But the drugs do not work as a vaccine would, by priming the immune system to actually prevent infection.

All this prompted the U.S. Centers for Disease Control to issue a series of guidelines to doctors considering prescribing PrEP regimens. Patients should be tested for existing infections and should undergo regular screening of kidney function, since that can be affected by the medication. Doctors should also get to know their patients well to determine how reliably they will take the medication and practice prudent sexual behavior.

GALLERY OF WELLNESS

For people already sick with AIDS, antiretrovirals can be lifesavers—as these images of patients in a South African program show. Ncapai Thobani (top) was at the brink of death in 1997. Now his viral load is undetectable in his blood. The woman in the middle photo chose to show her drugs, not her face, but is keeping her own improvement no secret. Her baby died of AIDS before the drugs were available. Pumla Dladla (bottom) and her daughter are both infected.

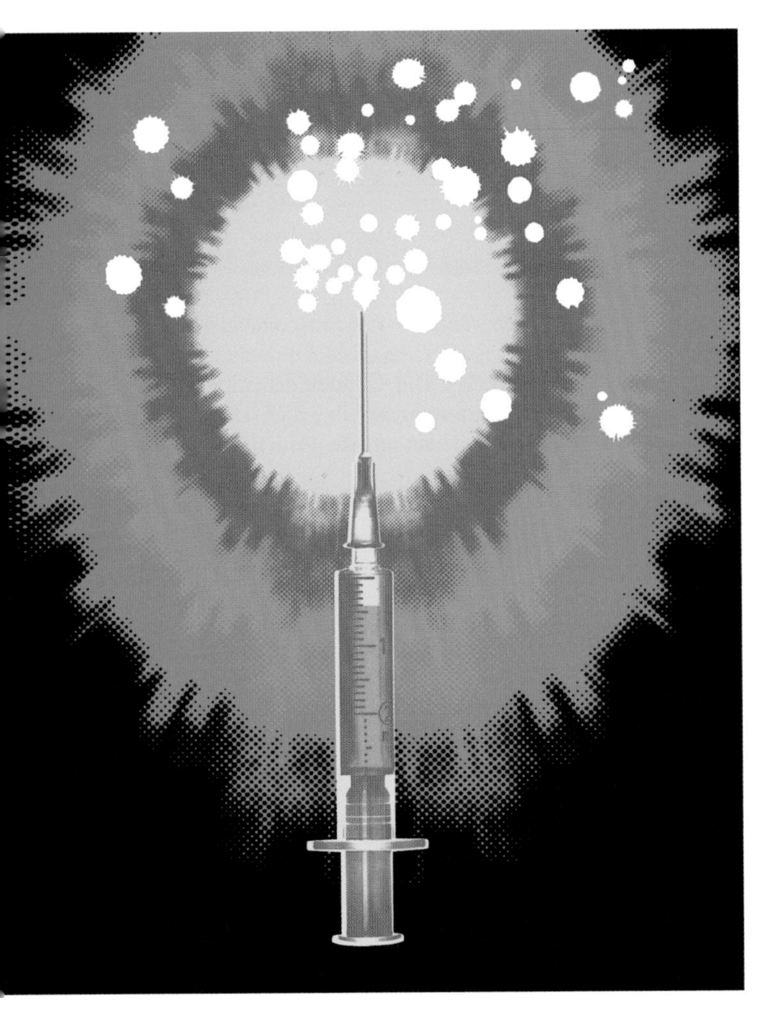

More Progress on Cancer Vaccines

Dr. Douglas Schwartzentruber and Dr. Larry Kwak would like to do what Dr. Jonas Salk once did: develop a vaccine that can help wipe out a disease. Nice idea, but here's the rub—the disease the two investigators are targeting is cancer, which in most cases is not caused by a pathogen. Without a virus, there's nothing for the vaccine to attack. That's the traditional thinking, but it's now being turned on its head.

Schwartzentruber, of the Goshen Center for Cancer Care at Indiana University, and Kwak, of the M.D. Anderson Center at the University of Texas, are pursuing similar goals in different labs. The idea is not to vaccinate healthy people against nonviral cancers they might never develop. Rather, vaccines would be used in patients whose disease has already been diagnosed and treated with surgery, chemotherapy, or radiation. The patients would then be immunized as a way to prevent the cancer from coming back and spreading. Schwartzentruber is studying melanoma, and Kwak, lymphoma, and they are taking aim at both diseases by exposing the immune systems of patients to concentrated proteins or other components of the cancers. That helps train the body to recognize the disease and battle back.

The experimental trials have their differences: Schwartzentruber uses tumor components common to all melanoma and adds a dash of Interleukin-2 as a booster. Kwak uses the patient's own cancer to make the vaccine. No matter. Both have had positive results in their ongoing trials—providing a promising new strategy against a very old disease.

New Ways Your Genes Make You Fat

The fight against obesity has engaged many fields of medicine: genetics to predict it; nutrition to prevent it; surgery to manage it; endocrinology to deal with the diabetes that often results from it. Enter Dr. Andrew Feinberg and Dr. M. Daniele Fallin, both at Johns Hopkins University, who are working on another contributor to obesity: epigenetics.

Epigenetics is the study of changes in gene activity that do not occur over generations, but rather in one person's lifetime, through experiences or exposures such as diet and environmental toxins. The changes occur in the cellular material that sits on top of the genome—the epigenome—and act as a dial, turning up and down the expression of various genes.

Feinberg and Fallin's paper, which involved 74 participants in Reykjavik, reports that certain patterns of epigenetic marks associated with 13 genes appeared to have a very clear relationship to body mass index (BMI). It's not yet possible to draw a direct cause-and-effect link between the genes and obesity, nor to say what environmental, dietary, or other factor caused the epigenetic changes in the first place. But once that answer is found, it may be possible to eliminate that factor in the lives of the already obese or at least prevent normal-weight people from ever encountering it.

EPIGENETICS IS THE STUDY OF GENETIC CHANGES THAT DO NOT INVOLVE MUTATIONS OVER GENERATIONS. RATHER, YOU ACQUIRE THEM DURING YOUR LIFE.

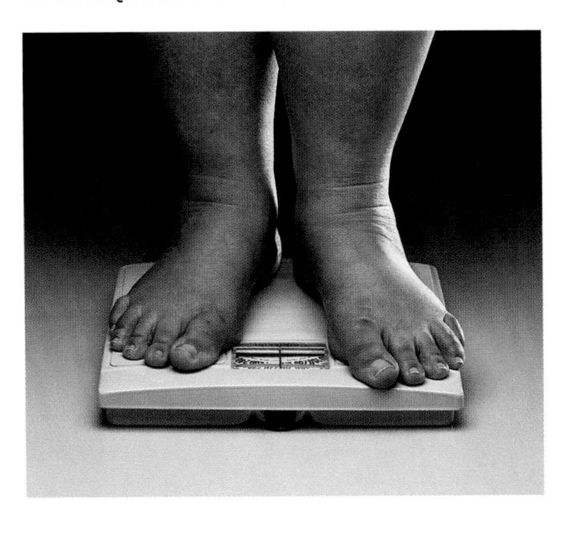

FROM TOP: AMY GUIP/GETTY IMAGES; JAMES W. PORTER/CORBIS

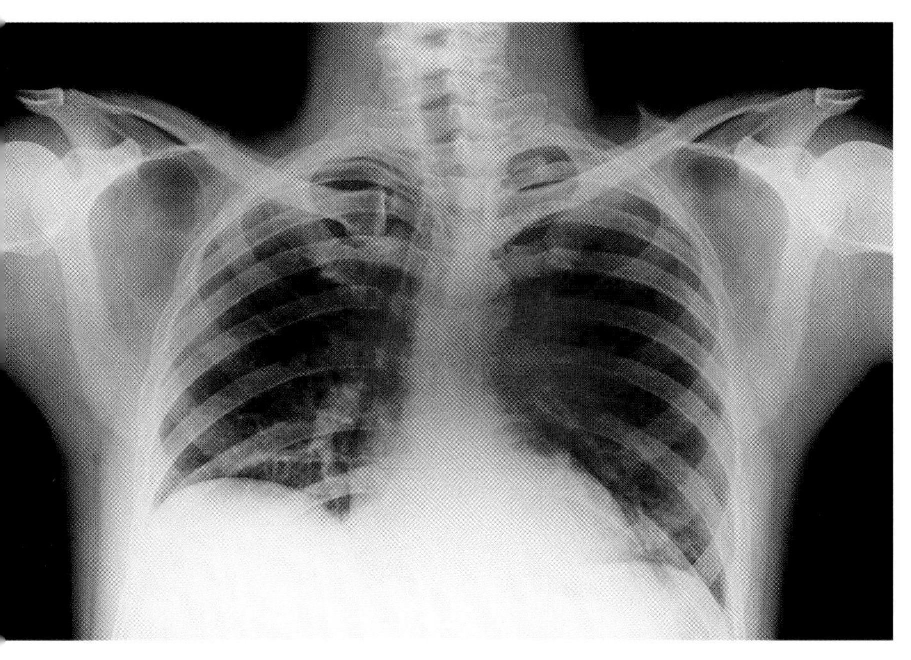

A Blood Test for Impending Heart Attack? Maybe

Heart attacks are frightening mostly because they are so unexpected. And while doctors have a range of tests that can predict who is at highest risk of having an event in the immediate future, those measures are not always sensitive and can be very invasive. So a group of experts decided to look at something easily accessible, like blood, for predictive factors. The idea was to find an alternative to the angiogram, in which technicians snake a long catheter from the large leg artery into the heart to view blockages.

The group tested a panel of 23 genes that make products detectable in the blood and found that such screening was 83% sensitive in picking up signs of obstructions typical of coronary artery disease. Using this test on top of a risk assessment that included clinical factors such as type of chest pain further improved the ability to predict which patients were at high or low risk. But those results, published in the *Annals of Internal Medicine*, are not enough to replace the angiogram quite yet, say some. "The whole concept of using genetic and protein risk prediction is a very important future-oriented strategy," says Dr. Ralph Sacco, president of the American Heart Association. "But I don't think in this case the test is ready for primetime."

Sacco also points out that other than angiograms, doctors can use less invasive tests, such as a stress test that measures how the heart responds under exercise or other kinds of strain, or a CT scan of heart vessels to check how well blood is flowing. "This opens up the intriguing idea that we can use genetic risk markers to better risk-stratify people," he says. "But they may be of more value in the future."

83%

SENSITIVITY OF THE BLOOD SCREEN IN PICKING UP SIGNS OF OBSTRUCTIONS TYPICAL OF CORONARY ARTERY DISEASE

IT JUST AIN'T SO ...

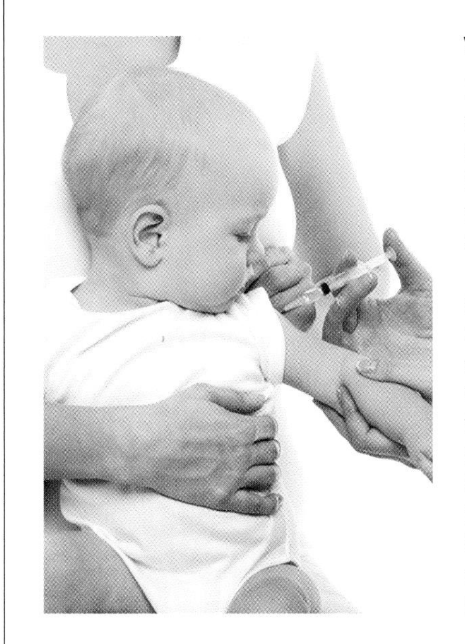

Vaccines Cause Autism! Oops, Never Mind

If there's a place nonsense goes to die, you can consign the idea that vaccines cause autism to that ignoble repository. The yarn began 12 years ago, when British medical journal *The Lancet* published a study indicating a link between autism and the mumps, measles, and rubella vaccine (MMR). The finding validated the fears of a growing subculture and played a role in the sharp drop in vaccination rates.

But a body of studies subsequently found no connection, and the final nail was hammered down in 2010, when *The Lancet* announced the official retraction of the study, finding the study's authors guilty of professional misconduct. The lead author, Dr. Andrew Wakefield, is charged not just with carrying out his work unethically—paying children at his son's birthday party to give blood and carrying out invasive tests such as colonoscopies and spinal taps on some subjects without regard for how they might be affected. He also failed to disclose that, while carrying out the research, he was being paid by lawyers acting for parents who believed their children had been harmed by the MMR jab. Despite this, some antivaccine activists might never be persuaded. Says Dr. Paul Offit, author of a book about the vaccine fight: "It's easy to scare people, but it's extremely hard to un-scare them."

FROM TOP: PETER DAZELEY/GETTY IMAGES; SCIENCE PHOTO LIBRARY/GETTY IMAGES

The Earth

ELECTRIC CARS ARE HERE ■ MONITORING FORESTS FROM ORBIT ■ BUILDING A GREENER ORANGE JUICE ■ DELICIOUS ALGAE ■ A SIMPLER KIND OF SOLAR CASH FROM CLEAN TECH ■ THIN-FILM ENERGY ■ A SMARTER GRID ■ THE FUTURE OF HIGH-SPEED RAIL

Saving the Earth by Reengineering It

If human carelessness can damage the planet, can human engineering fix it? The future of our climate may turn on the answer.

BY BRYAN WALSH

If the world is going to come to grips with the climate change crisis, we must drastically reduce greenhouse gas emissions in the decade ahead, a challenge that will require remaking the way we produce and use energy at a likely cost of trillions of dollars. Or maybe we can just change the color of the sky: Former Microsoft CTO Nathan Myhrvold, now at Intellectual Ventures, a Seattle-based venture capital firm and think tank established by Microsoft co-founder Paul Allen, has devised a plan to pump 100,000 metric tons of sulfur aerosol particles into the atmosphere every year. Sprayed into the stratosphere using 18-mile-high, two-inch-wide tubes held in place by balloons, the sulfur particles would scatter incoming sunlight, bouncing some of it back into space and reducing the amount that reaches Earth's surface. Computer models indicate that Myhrvold's method, carried out indefinitely, would probably keep the planet cool even if carbon emissions continued growing, though there would be some side effects—like a yellower, hazier sky. Still, the system would cost just $20 million to set up and $10 million a year to run—a minuscule fraction of the likely price of transforming the global energy system.

Myhrvold's proposal, which he has called the StratoShield, is an example of geoengineering: an attempt to control the climate directly. A concept long kept in the shadows—considered too radical to try—geoengineering is gradually becoming more and more prominent in climate-change discussions. That's in part a reflection of the world's near-total failure to curb carbon emissions, even as global temperatures continue to rise and climate models warn of an ever more frightening future. Geoengineering advocates present their concepts as the planet's ultimate plan B, a backup that could be

FRED HIRSCHMANN/SCIENCE FACTION/CORBIS

GOING, GOING ... *A blue ice cave in Alaska's Glacier Bay National Park is being hollowed out by rising global temperatures.*

STASHING THE CARBON *The seas and trees are natural carbon sinks. Plankton blooms could gobble up still more, as could expanded forests.*

deployed quickly if climate change really began to get out of hand. But while the economics of geoengineering are tempting, even those who work in the field worry that the existence of an emergency button might sap the will of governments to do the hard but necessary work of reducing emissions. And although climate models are improving, we may still know far too little about the planet to begin messing with the global thermostat—which could produce side effects as dangerous as the problem itself. "We are overdue to research this," says David Victor, director of Stanford University's Energy and Sustainable Development Program and an advocate of geoengineering work. "But there needs to be a very high bar for how much care we take."

If you think trying to craft a global deal to control carbon emissions has been tricky, trying to get 192 independent countries to agree on how to reengineer the planet could be all but impossible.

Geoengineering breaks down into two broad categories: direct carbon reduction (CDR)—taking carbon out of the atmosphere and storing it; and solar radiation management (SRM)—artificially blocking sunlight to cool the planet. The first, CDR, is already a part of our climate-management schemes—forests, after all, suck in and store billions of tons of carbon, as does the deep ocean. Most of the CDR-based geoengineering techniques being developed—like "artificial trees" that chemically capture CO_2 directly from the air and store it—shouldn't produce unexpected side effects. Another method—seeding the ocean with iron to encourage plankton blooms, which will then absorb carbon—does worry activists, who fear the impact on marine systems. Broadly, though, CDR doesn't scare too many people. "Conceptually it's all the same as planting a forest," says Ken Caldeira, an atmospheric researcher at the Carnegie Institution for Science at Stanford.

Cost is the main limitation on CDR techniques. Agreements are already in place to try to curb deforestation for

FROM LEFT: IAN SHIVE/AURORA PHOTOS; FABIAN GONZALES/ALAMY

Controlling the Climate

DIRECT CARBON REDUCTION (DCR): The simplest way to curb climate change is to pull excess carbon out of the atmosphere. Forests do it naturally, and so-called artificial trees—synthetic carbon scrubbers—could build on that idea.

SOLAR RADIATION MANAGEMENT (SRM): Limiting the amount of sunlight that reaches the surface can keep the planet cooler. Seawater or sulfur sprayed into the sky could do that, as could wilder schemes—like orbiting giant mirrors.

IRON SEEDING: One form of carbon reduction that gets a lot of attention entails seeding the ocean with iron to encourage carbon-eating plankton blooms. The plan would probably work, but it could upset the balance of marine life.

climate reasons, but they've been only partially successful because the market for wood products and the drive to clear land for agriculture is so strong. Caldeira believes that the only way CDR would be feasible is through a carbon price—a high one, close to $100 a ton—that would make it financially viable to start doing things differently.

Advocates of SRM see a different set of possibilities and challenges. Their ideas range from the ambitious—constructing a gigantic suborbital mirror to bounce back sunlight—to the innocuous, like painting highways and roofs white to reflect sunlight rather than absorb it. But the most viable SRM techniques involve plans like Myhrvold's, to pump sulfur—or something—into the atmosphere.

Scientists know that blocking some of the sunlight that reaches Earth does lower temperatures. When the volcano Mount Pinatubo exploded in the Philippines in 1991, spewing millions of tons of sulfur into the atmosphere, global temperatures dropped 1°F in less than a year. Volcanic ash is unpredictable, but another SRM method would involve ships spraying nothing but seawater into the sky, brightening marine clouds and enhancing their reflectivity.

Even if SRM engineering were successful, it's not something you could do once and forget about it. Seeding or spraying would have to continue as global carbon levels rose, and if you stopped because of a war or economic crisis, a bounce-back effect could cook the planet in a matter of years. Geoengineering also wouldn't offset some of the other effects of high carbon emissions, like ocean acidification. What's more, the acid rain that would come from sulfur seeding would present problems of its own. If Earth were a human patient, geoengineers would be like 19th-century doctors, who in their ignorance sometimes did more harm than good. And if you think trying to craft a global deal to curb carbon emissions has been tricky, trying to get 192 nations to agree on exactly how we should fine-tune the climate could be all but impossible. "You are not going to get a reasonable global dialogue about a geoengineering research program," says Victor.

Still, the possible consequences of unchecked global warming are so scary that it would be smart to explore geoengineering so that at least we have an idea what will happen if we push the doomsday button. Little official work is being done outside computer models, however, and while a few tycoons like Bill Gates have funded small-scale geoengineering research, no government has yet done so. The legal picture is murky too. Last year the Convention on Biological Diversity (CBD) essentially banned geoengineering research except for small-scale trials, but no one knows exactly what qualifies and what doesn't. (And in any case the U.S. isn't a party to the CBD.) Scientists worry that in the absence of clear rules and government funding, geoengineering might fall to rich individuals or rogue nations, unbound by ethics. "There is a race between responsible geoengineering research and those who will do it unilaterally," says Victor. "I'm worried countries could go off and just do things on their own."

The right approach would be to research geoengineering in a fully open manner—and hope we never need to use it. But the reality is that, like nuclear weapons, the possibility of geoengineering can't be put back in the bottle, as much as we might wish we could. "We are in the [planetary] gardening business, whether we want to be or not," says David Keith, a climate scientist at the University of Calgary and one of the world's leading experts on geoengineering. "The question is whether we'll be responsible gardeners or irresponsible gardeners." Only the fate of the planet—and all of us—will depend on the answer.

2050

BY THAT YEAR, WIDESPREAD ADOPTION OF PLUG-INS COULD REDUCE GREENHOUSE GAS EMISSIONS BY 450 MILLION METRIC TONS ANNUALLY—EQUIVALENT TO REMOVING 82.5 MILLION PASSENGER CARS FROM THE ROAD.

Cars Go Electric, and the Roads Go Green

The electric car, so long promised, may finally be pulling into your driveway. In the U.S., General Motors recently introduced the plug-in Volt, which should, if it lives up to the hype, get a tidy 230 miles a gallon in ordinary driving. Daimler is doing a test run on an electric version of its baby Smart car and claims to get the equivalent of 300 mpg. Nissan is offering its all-electric Leaf, advertised at a dazzling 367 mpg. Meantime, China—the once-slumbering giant in the electric car game—is stirring at last. Beijing knows that promoting electric vehicles could be a way to stem the country's rising dependence on imported oil and clear its polluted air. Foreign automakers may have a century-long headstart on conventional cars, but Chinese companies can compete on new electric technology today—particularly with the help of surging domestic companies like auto manufacturer Coda and battery makers Lishen, based in Tianjin, and BYD, in Shenzhen.

America, once the global leader in all things automotive, would like to reclaim that title, but the planet will be the real winner regardless of which country emerges first in electrics. Electricity is far cheaper than the cheapest oil—plug-ins generally run on the equivalent of 75 cents a gallon. Even with America's current electrical supply, which is more than 50% coal-generated, switching to plug-ins will reduce greenhouse gases, and as the grid gets cleaner, those savings will increase. A study by the Pacific National Northwest Laboratory found that the grid could power 73% of the nation's car fleet without adding a single new plant, provided most of the charging was done at night, when demand is lowest. The age of fossil fuels—and fossil emissions—may at last be ending.

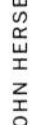

Keeping an Eye on the Forests From Space

Google never quits. The company that revolutionized the search engine and e-mail is now taking users where only astronauts have gone before. At the 2010 UN climate summit in Cancún, the company announced the creation of the Google Earth Engine, a new technology platform that will enable global monitoring of change in the planet's environment. Google has tapped a quarter-century of satellite images provided by Landsat, along with data that includes MODIS, a major weather-tracking project, to create the new eye in the sky. The purpose of the project, as chief engineer Rebecca Moore put it, is to create "a living, breathing model of the Earth with all of the data and analysis that's available."

The data Earth Engine will tap isn't new, but Google will make it far more accessible and far more searchable than it has ever been before. That will be a major boon for environmental researchers, for whom information is their lifeblood. And it's particularly important for projects on preventing deforestation—vital work, since loss of first acreage is responsible for 12% to 18% of annual greenhouse gas emissions. If researchers can't track the rate at which trees are being lost over time in a country like, say, Bolivia, it's impossible to design an effective scheme to fix the problem that will be both fair for the people and good for the climate. Forestry is one of the few bright spots in environmentalism right now—new loss of acreage in the Brazilian Amazon has dropped to a record low—and technology could make it brighter.

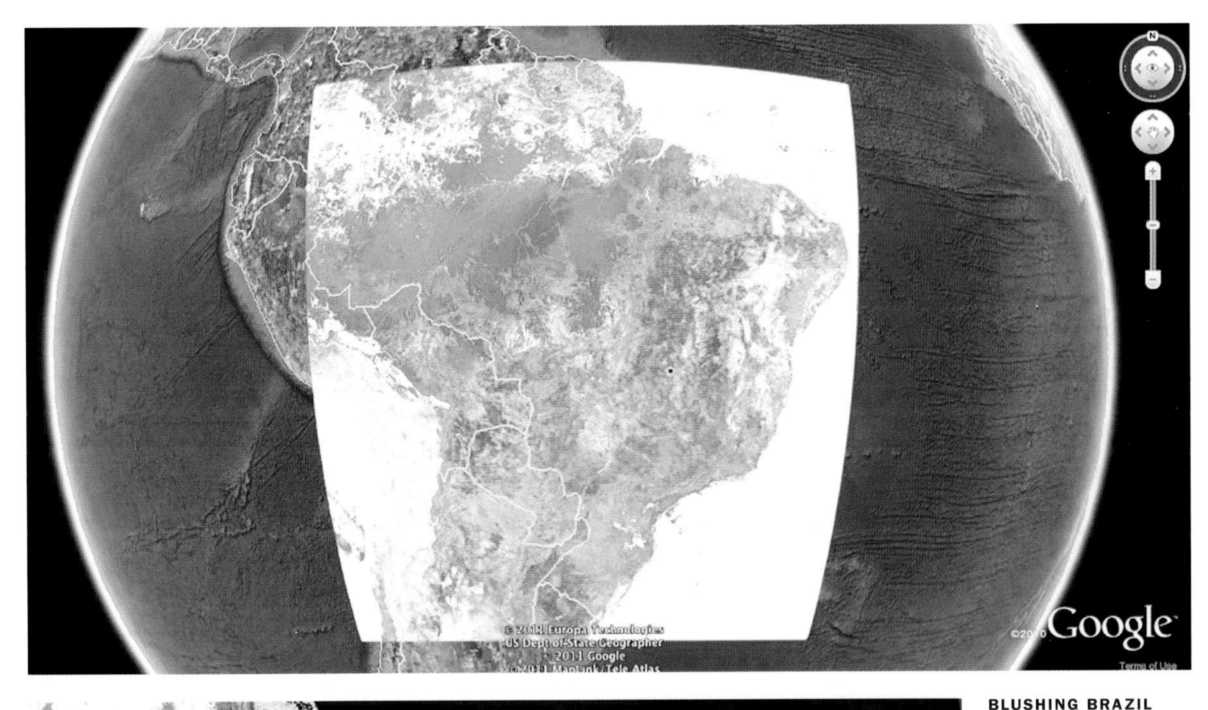

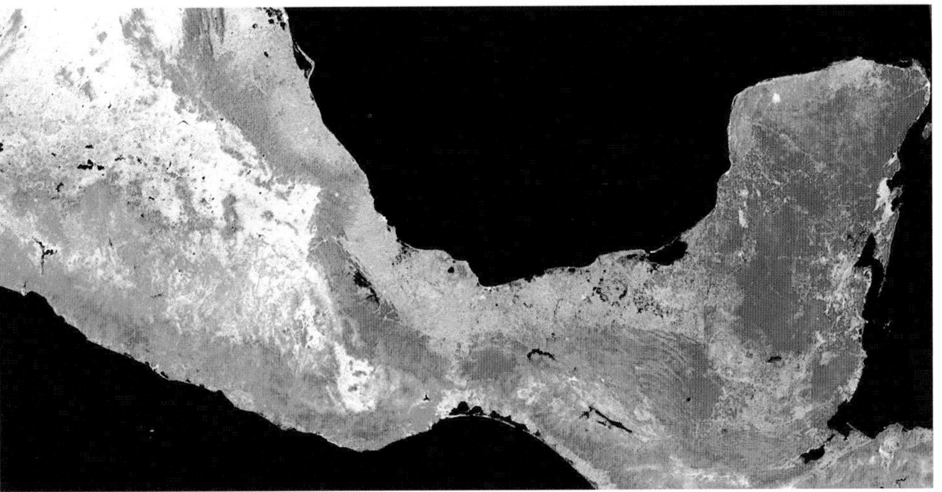

BLUSHING BRAZIL
Google Earth Engine gives the Amazon a color-coded overlay (above) to make the health of the jungle easy to read. The different shades represent green vegetation, non-photosynthetic vegetation, and bare soil cover. Images taken over time will allow environmental monitors to track both deforestation and slower forest degradation. The map of southern Mexico and the Yucatán (left) uses a different system to survey forest cover and water flow.

FROM TOP: C. SOUZA/IMAZON/GOOGLE; M.HANSEN/SDSU/CONAFOR/GOOGLE

FERTILIZER ACCOUNTS FOR ABOUT 35% OF THE CARBON FOOTPRINT IN OJ—MORE THAN THE FUEL USED SHIPPING IT TO STORES.

Tropicana: Trying to Make a Kinder, Gentler Orange Juice

How green is your orange juice? A couple of years ago PepsiCo, which owns the Tropicana brand, tried to size up the carbon footprint of the popular morning tonic. It found that each half-gallon carton of OJ is responsible for 3.75 pounds of CO_2. The single biggest contributor is the fertilizer used to grow the orange trees.

What was needed was a greener fertilizer—and that's exactly what the company is experimenting with. Working with agricultural companies Yara International and Colorado-based Outlook Resources, PepsiCo will test low-carbon fertilizers at one of its producer farms in Bradenton, Fla. Most fertilizer is made with nitrogen, which in turn releases a lot of nitrous oxide, a greenhouse gas. Yara is experimenting with calcium-based fertilizer that would cut nitrous oxide emissions almost completely. The fertilizer-manufacturing process also relies on natural gas as fuel, another source of carbon. Yara is looking to improve the efficiency of its manufacturing plants—but Outlook is going even further: The company hopes to eschew imported natural gas entirely in favor of organic, locally sourced biofuels. The local sourcing helps cut the carbon emissions associated with transportation, while the use of alternative fuels cuts carbon emissions further. More than 13 million tons of fertilizer are used on U.S. farms a year. A less carbon-intensive way to help plants grow could mean a less carbon-heavy atmosphere.

Cookies Made With Yummy Scum

The most unexpected thing about the next diet foods to hit your plate may be their secret ingredient: all-natural algae—those single-celled organisms that float on stagnant ponds. And it's not an agribusiness giant or some boutique bistro that will be responsible. Your meal will come courtesy of Solazyme, a biofuel startup in South San Francisco.

Solazyme, like a number of other 21st-century companies, has been working on developing algae as a building block of green gasoline, but along the way the in-house chemists discovered that an algae-derived flour could be substituted for conventional flour to make a product similar to pita bread. The lipid profile of the algae flour was strikingly similar to that of healthy olive oil. Using the algae-derived flour, they have been able to make cookies, snack drinks, dipping sauces, and more—all with less fat and calories and more protein than conventional food.

Solazyme isn't likely to sell its cookies or drinks directly; rather, it will produce the raw algae flour and sell it in bulk to food manufacturers. The challenge will be convincing consumers that algae is worth eating. Although folks in other parts of the world, including Southeast Asia, have long been accustomed to algae as a traditional food ingredient, Americans may well balk at the idea of eating cakes and cookies made with what is, after all, pond scum. But if Solazyme can help make low-fat foods that really taste like dessert, hungry dieters will be lining up to eat their algae.

CLOCKWISE FROM TOP RIGHT: JACOB KEPLER/BLOOMBERG/GETTY IMAGES; DAVID NUNUK/VISUALS UNLIMITED; SOLAZYME; GETTY IMAGES

Solar Power Done a Whole Different Way

Smack in the middle of the Mojave Desert, the Las Vegas area gets around 330 days of sunshine a year. That's good news for the Spanish renewable-power company Acciona Energy. Acciona is the developer of Nevada Solar One (NS1), which is solar power with a twist: harnessing the heat of the sun, not just its light. Instead of directly converting sunlight into electricity with photovoltaic panels—the kind you might see on rooftops—NS1's solar thermal technology uses rows of specially curved parabolic mirrors to focus sunlight on a pipe full of synthetic oil. The sun's energy superheats the oil, which is then used to boil water into steam. The steam runs turbines, which generate electricity.

The technology is as simple as any fossil fuel plant, and cheaper than technologically complex photovoltaic panels. It can also be more easily built up to utility scale than photovoltaic solar. Acciona's plant, which began operation last year, produces 64 megawatts of electricity for the utility company Nevada Power, enough to light up 14,000 homes. The company's Spanish competitor Abengoa just announced a plan to build a 280-megawatt solar thermal plant outside Phoenix, which would be the largest such project in the world.

Size is everything when it comes to solar thermal. NS1 uses 182,000 parabolic mirrors spread over 400 acres of flat desert, creating a glistening sea of glass visible from miles away. Up close the mirrors are shaped like shallow satellite dishes, chasing the sun's movement as it passes through the sky. Although the plant might look fragile, it's not; NS1 has lost only 10 mirrors in nine months of operation.

Acciona is planning on adding 500 megawatts of solar thermal power in the U.S. by the end of 2011. That's just a fraction of the new power capacity a growing America will need over the coming years, but there's a chance that with the right federal policies, solar thermal could contribute far more. Its proponents believe that alone among major renewables, solar thermal has the capacity to displace fossil fuels on the utility scale, perhaps eventually providing a quarter or even half of the national power supply.

HEATING UP *Two views of the Nevada Solar One array. More than 182,000 mirrors are spread across 400 acres, producing enough power for 14,000 homes.*

BY 2022, WASHINGTON WANTS
THE U.S. TO BE PRODUCING
36 BILLION GALLONS
OF RENEWABLE FUELS A YEAR, UP FROM
11.1 BILLION TODAY. THERE'S A LOT OF
MONEY TO BE MADE IN THAT EFFORT.

FROM LEFT: CJ BURTON/CORBIS; JOHN MOORE/GETTY IMAGES

Strange Bedfellows in the Clean-Tech Sector

If we're going to find a way to fix our long-term energy woes, the solution is likely to come from Northern California. Yes, in Silicon Valley the same entrepreneurs who brought us the Internet are exploring new ways to make and use energy. There's gold as well as good in the effort. The research company Cleantech Group estimates that by 2020, the global clean-tech sector will be worth more than $3 trillion and could account for as much as 15% of some nations' GDP.

The problem is that clean-tech startups run on venture capital—and VC money, like just about every other form of financing, fell off a cliff during the recession, dropping 33% in 2009. Not to mention that creating a new energy company is much more challenging than building, say, a major dotcom player, because energy companies often need lots of capital to finance manufacturing. But help could be coming from the giant, slow-moving corporations that many clean-tech startups are trying to replace. Here, too, wealth is the goal: Those established manufacturers will be looking to tap the next great revenue stream by snapping up—or forming joint ventures with—the Silicon startups.

Those changes are already beginning to happen. For example, the $50 billion French company Veolia has announced the launch of the Veolia Innovation Accelerator, which will seek to build partnerships with new clean-tech companies. It will start with NanoH2O, a startup that has pioneered membranes for use in desalinization plants, which can make it less expensive to produce clean water.

The program is a tacit admission that even the best corporations need to go outside to find smart ideas—and even the spunkiest startups may have to go old-school. But the partnership isn't without risk. As one venture capitalist put it, startups can suffer "death by pilot project," in which a good idea gets caught in the institutional eddies of a major corporation. Still, in an energy-hungry world, the stakes are too high for the two sides not to find a way to get along.

Solar's New Look: Thin Film

There's an unmistakable clunkiness to traditional solar panels—the heavy, sun-catching frames that haven't changed much since Jimmy Carter ordered some installed on the roof of the White House in the 1970s. Or they *hadn't* changed much. But solar has gone on a diet. The newest photovoltaic innovation is not panels, but sheets—filmy, flexible layers of indium, gallium, and diselenide that turn sunlight into electricity.

Thin-film solar is a product that's all at once hot—and cheap—throughout the solar world, from Tucson to Silicon Valley to Germany. Though film is cheaper than the crystalline panels on most rooftop assemblies, the technology had proved maddeningly difficult to mass-produce, which had kept it from going mainstream. But today thin film is the hottest part of the fastest-growing new energy source in the world. BCC Research, which charts technology markets, expects the global solar market to grow from $13 billion to $32 billion by 2012, with thin film expanding 45% a year. Masdar, the clean-energy arm of the government of Abu Dhabi, announced that it will invest $2 billion in thin film.

The key is new manufacturing technology. The multinational Nanosolar, for example, achieves radical cost savings by directly applying photoactive chemicals with an ink composed of nanoparticles. The goal: to sell the film profitably at $1 a watt. At that point solar becomes cheaper than coal. Nanosolar says it will be there soon. If so, it will turn the energy community upside down.

The Grid Gets Smarter and Your Bills Get Smaller

In 1886 the town of Great Barrington, Mass., set up the first alternating-current electric transmission line in the U.S. In the 125 years since, the national grid it gave rise to has changed little. The result is a creaky system that is still prone to spectacular failures, like the 2003 blackout in the Northeastern U.S. and parts of Canada. Even on its best days the grid is inefficient, with 7% to 9% of power lost in transmission. Energy companies have little way of tracking the amount of power they produce and distribute, and households have little way of monitoring how much they consume or what they're paying for it—at least until the bill comes.

But there's a way to upgrade the grid by marrying the networked intelligence of the Internet to transmission lines and transformers. The result wouldn't just be better, it'd be smarter—a smart grid. Utilities would be able to remotely monitor power use, allowing them to respond rapidly to outages. Consumers would be able to use intelligent, networked appliances to control how and when they use electricity, shrinking their power bills and smoothing demand.

A smarter grid could better integrate intermittent renewable sources like wind and solar, which would help cut carbon emissions and ultimately save consumers as much as $20 billion over the next decade. Though transforming the nation's electrical system will be a long and expensive process, it is one of the White House's top green priorities; the federal government released $3.4 billion in grants in 2009 to 100 companies working on the grid.

An electric grid that works like the Internet could have a downside. There are concerns that it could be a tempting target for hackers or even terrorists, though similar fears haven't stopped the growth of e-commerce. What's more, e-commerce only makes shopping easier. A smart grid makes the whole planet healthier.

7–9%

HOW MUCH POWER IS LOST IN THE WIRES OF OUR TRANSMISSION SYSTEMS

HOW A SMART HOME WILL WORK ON A SMART GRID

THERMOSTAT Your heat and AC could surf the wave of local electricity use, dialing up when demand is lowest and down when it's highest. The override switch, of course, remains in your hands.

PLUG-IN CAR Smart cars could be plugged into smart grids, drawing electricity when it's most affordable and even selling some back when the price is right.

DISHWASHER Tracking electricity costs throughout the day, the unit could wait until washing the dishes would cost you the least and then switch on.

LIGHTING Lamps and other light fixtures can already switch themselves off when no one's in the room or the ambient lighting is sufficient. Now they could respond to price signals too.

COMMAND CENTER Homes will be equipped with data dashboards, allowing you to monitor every watt you use and tweak your consumption accordingly.

GETTY IMAGES

Hope for High-Speed Rail

Let's be honest: America's railroads are a mess. They're a laughingstock compared with those in other developed nations and, increasingly, even in developing nations like China, which is investing more than $300 billion to build more than 16,000 miles of high-speed track by 2020. Today you can travel the 250 miles from Paris to Lyon on the high-speed TGV in two hours. Covering a similar distance from Philadelphia to Boston takes some five hours, and that's on an Amtrak Acela train, the closest thing we have to high-speed rail.

Oddly, Americans seem resistant to changing things. When Washington released $8 billion in stimulus money to states to build high-speed rail, some, like Florida, said no thanks. The problem for many is that no one pretends the bill would stop at $8 billion. Even Vice President Biden, a very vocal cheerleader for rail, called that initial outlay "seed money."

Nevertheless, high-speed rail is an idea whose time has come—and not just for travelers weary of both the ordeal and expense of air travel. According to Environment America, high-speed rail uses a third less energy per mile than auto or air travel, and a nationwide system could reduce oil use by 125 million barrels a year, doing the carbon-choked atmosphere a favor. In addition, high-speed rail represents the kind of long-term infrastructure investment that will pay back for decades, just as the interstate highway system of the 1950s has.

FROM LEFT: DOUGLAS GRAHAM/ROLL CALL/GETTY IMAGES; GENT SHKULLAKU/AFP/GETTY IMAGES

IT JUST AIN'T SO ...

APRÈS LE DÉLUGE *A flooded street in Shkodra, Albania*

Climate Change a Hoax? Please!

Let's pretend we didn't want to believe that dumping industrial chemicals into rivers was bad for fish. That's an easier case to make than you'd think. You can't really see the fish swimming beneath the surface, so how do you know? The water still looks blue, and if fishermen report that their hooks have been coming up empty—well, maybe that's their own fault. Repeat that enough and you can pretend there's a controversy, and where there's controversy, there's uncertainty. No need to take drastic action until all the facts are in.

That's the kind of fiddling we're still doing as our overheated, greenhouse-gassed Earth continues to burn. The numbers are numbingly familiar by now—the 34 consecutive years global temperatures have been above average, the nine years since 2000 that have ranked in the top 10 hottest ever, the 49 states that all had snow on the ground at one time in the U.S. in 2011. Climate change is a crazy time for the atmosphere—floods here, droughts somewhere else; record blizzards here, record hurricanes there. Just as a high fever can cause you to sweat and shiver simultaneously, so, too, does the planet go haywire when its thermostat comes unsprung.

All this has been in the climate models for decades, and it all is unfolding as predicted. Yes, the models are imperfect, and yes, the modelers themselves have on one or two occasions played cute with their data. But the scientific truth is still clear. Climate-change deniers are motivated by a lot of things—some perfectly legitimate. No one wants to take draconian steps that aren't necessary. But if our doctor had been warning us for the past 30 years that we were on our way to becoming gravely ill and now we were developing all the symptoms, we'd be fools to keep saying, "Prove I'm sick." Wisdom would lie in asking, "How do I get well?" Let's resolve to be wise.

Zoology

ELEPHANT SEALS AS RESEARCH ASSISTANTS ■ THE SEXUAL DECEIT OF ANTELOPES ■ BIO-EXUBERANCE IN THE SOUTH PACIFIC ■ FIGHTING PTSD WITH DOGS ■ INVASION OF THE PYTHONS ■ A FLU-PROOF CHICKEN THE TALIBAN'S TERRORIST MONKEYS ■ THE EPIDEMIC KILLING BEES ■ ANIMAL SUICIDE ■ YOUR DOG THE GENIUS

Inside the Minds of Animals

Science is revealing just how smart other species can be—and raising questions about how we treat them.

By Jeffrey Kluger

If you visit the Great Ape Trust in Des Moines, give some thought in advance to what you'd like to say to Kanzi (at right). Like most bonobos, a close but more peaceable cousin of the chimpanzee, Kanzi has a very loud and serviceable voice, although it's not especially good for forming words. But that doesn't mean he isn't talkative.

For much of his day Kanzi keeps a sort of glossary close at hand—three laminated, placemat-like sheets filled with hundreds of colorful symbols that represent all 384 words he's been taught by his minders or picked up on his own. He can build thoughts and sentences, even conjugate, all by pointing. The sheets include not just easy nouns and verbs like *ball* and *Jell-O* and *run* and *tickle* but also concept words like *from* and *later* and *to* and *before.* The not-for-profit Trust is home to seven bonobos, including Kanzi's baby son, Teco, born in June 2010, all of which are being raised from birth with language as a constant feature of their days. On one recent morning Kanzi used his glossary sheet to indicate that he wanted to play with his ball. He waited patiently while someone went to find it. When the ball was finally brought to him, an attendant asked, "Are you ready to play?" Kanzi looked up balefully. "Past ready," he pecked.

Humans have a fraught relationship with beasts. They are our companions and our chattel, our family members and our laborers, our household pets and our household pests. We love them and cage them, admire them and abuse them. And, of course, we cook and eat them. Our dodge—a not unreasonable one—has always been that animals are ours to do with as we please simply because they don't suffer the way we do. They don't think, they don't worry. They may pair-bond, but they don't love. "The reason animals do not

FINLAY MACKAY

speak as we do is not that they lack the organs," René Descartes once said, "but that they have no thoughts."

Yet one by one the berms we've built between ourselves and beasts are being washed away. Humans are the only animals that use tools, we used to say. But what about the birds and apes we now know do as well? Humans are the only ones empathic and generous, then. But what about the elephants that mourn their dead and the rats that react to the pain of another rat? As for humans being the only beasts with language? Kanzi himself could tell you that's not true.

There are a lot of obstacles in the way of our understanding animal intelligence—not the least being that we can't even agree whether nonhuman species are conscious. We accept that chimps and dolphins experience awareness; we like to think that dogs and cats do. But what about mice and newts? What about a fly? Is anything going on there at all? There's more than species chauvinism in that question.

"Below a certain threshold, it's quite possible there's no subjective experience," says cognitive psychologist Dedre Gentner of Northwestern University. "I don't know that you need to ascribe anything more to the behavior of a cockroach than a set of local reflexes that make it run away from bad things and toward good things."

Where the line should be drawn is impossible to say. Still, most scientists agree that awareness is probably controlled by a sort of cognitive rheostat, with consciousness burning brightest in humans and other high animals and fading to a flicker—and finally blackness—in very low ones.

Among animals aware of their existence, intellect falls on a sliding scale as well, one often seen as a function of brain size. Here humans like to think they're kings. The human brain is a big one—about three pounds. But the dolphin brain weighs up to 3.75 pounds, and the killer whale carries a monster-size 12.3-pound brain. Still, we're smaller than dolphins and much smaller than whales, so correcting for body size, we're back in first, right? Nope. The brain of the Etruscan shrew weighs just 0.0035 ounce, yet relative to its tiny body, its brain is bigger than ours.

While the size of the brain certainly has some relation to smarts, much more can be learned from its structure. Higher thinking takes place in the cerebral cortex, the most evolved region of the brain and one many animals lack.

Crows and other corvids excel at tool use, a function of clever neurology that allows simple brain structures to multitask.

Mammals are members of the cerebral-cortex club, and as a rule, the bigger and more complex that brain region is, the more intelligent the animal. But it's not the only route to creative thinking. Consider tool use. Humans are magicians with tools, apes dabble in them, and otters have mastered the task of smashing mollusks with rocks to get the meat inside. But if creativity lives in the cerebral cortex, why are corvids, the class of birds that includes crows and jays, better tool users than nearly all other nonhuman species?

Crows, for example, have proved themselves adept at bending wire to create a hook so that they can fish a basket of food from the bottom of a plastic tube. How the birds perform such stunts without a cerebral cortex probably has

Who's the Smartest?

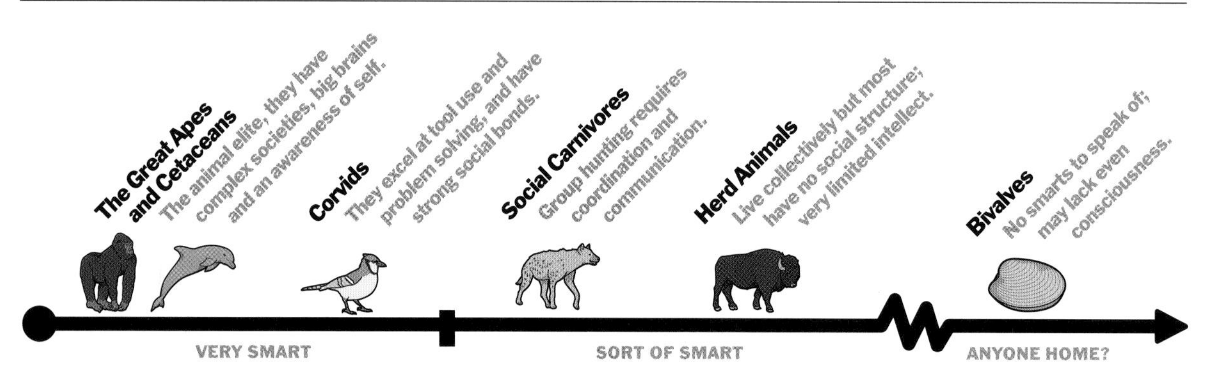

CLOCKWISE FROM TOP: FINLAY MACKAY; MICHAEL PATRICK O'LEARY/CORBIS; ILLUSTRATION BY BROWN BIRD DESIGN

something to do with a brain region they do share with mammals: the basal ganglia, more primitive structures involved in learning. Mammalian basal ganglia are made up of a number of parts, while those in birds are streamlined down to one. Earlier this year a collaborative team at MIT and Hebrew University of Jerusalem found that while the specialized cells in each section of mammalian basal ganglia do equally specialized work, the undifferentiated ones in birds' brains multitask, doing all those jobs at once.

It's easy enough to study the brain and behavior of an animal, but subtler cognitive abilities are harder to map. One of the most important skills human children must learn is something called the theory of mind: the idea that not all knowledge is universal knowledge. A toddler who watches a babysitter hide a toy in a room will assume that anyone who walks in afterward knows where the toy is too.

The theory of mind—knowing that what's in my head is different from what's in yours—is central to communication and self-awareness, and it's the rare animal that exhibits it, though some do. Dogs understand innately what pointing means: that someone has information to share and that your attention is being drawn to it so that you can learn too. That seems simple, but only because we're born with the ability and, by the way, have fingers with which to do the pointing.

Great apes, despite their impressive intellect and five-fingered hands, do not seem to come factory-loaded for pointing. But they may just lack the opportunity to practice it. A baby ape rarely lets go of its mother, clinging to her abdomen as she knuckle-walks from place to place. But Kanzi, who was raised in captivity, was often carried in human arms, and that left his hands free for communication. By the time he was 9 months old, he was already pointing at things.

Pointing isn't the only sign of a species that grasps the theory of mind. Blue jays—another corvid—cache food for later retrieval and are mindful of whether other animals are around to witness where they've hidden a stash. If the jays have indeed been watched, they'll wait until the other animal leaves and then move the food. Some animals also grasp abstractions such as sameness and difference—shown in tests in which they distinguish between a picture in which two objects match and another in which they don't.

If animals can reason—even if it's in a way we'd consider crude—the unavoidable question becomes, Can they feel? Do they experience empathy or compassion? It's well established that elephants seem to grieve, lingering over a herd mate's body with what looks like sorrow. They show similar interest—even apparent respect—when they encounter elephant bones. Empathy for living members of the same species is not unheard of either. When rats are in pain and wriggling, other rats that are watching will wriggle in parallel, suggesting that they are suffering by proxy.

There will never be any question that humans are overwhelmingly the planet's smartest species, but that does not mean that other species aren't remarkable in their own way. Ultimately, the same biological knob that adjusts animal consciousness up or down ought to govern how we value the way those species experience their lives. Kanzi's glossary is full of words like *noodles* and *sugar* and *candy* and *night*, but scattered among them are also *good* and *happy* and *be* and *tomorrow.* If it's true that all those words have meaning to him, then the life he lives—and by extension, those of other animals—may be rich and worthy ones indeed.

Elephants are very familial and appear to mourn their dead. If they find elephant bones, they gently examine the skull.

Enlisting Elephant Seals as Research Assistants

The sea doesn't give up its secrets easily. Water may cover 70% of the planet, but unless you get inside it, you don't know a lick about what's going on. Now the National Oceanic and Atmospheric Administration (NOAA) has come up with an imaginative way to monitor stocks of one of the world's favorite and most beleaguered species of fish: salmon. The trick? Enlist seals as census takers.

Scientists tracking the comings and goings of salmon typically capture a few individuals and tag them with electronic transmitters—a strategy that's simple but limited. Transmitters need receivers, and they have generally been installed in fixed spots like ocean shelves. If the fish don't happen to pass by, you get no data at all.

NOAA scientists realized that the answer was to draft the elephant seal: Glue a transponder to the hide of the animal, release it back into the wild, and the odds are pretty good it will always be near some salmon, since the two species favor the same kind of waters and have similar traveling patterns.

If it succeeds, it could easily expand. Commercial and military vessels already traveling on their own routes could carry the same kind of receivers, and as the transmitting tags get smaller and cheaper, they could be put on a greater number of fish and mammals.

How Faking Fear Helps Topis Mate

Male animals hoping to mate know that displaying strength and courage is the best way to attract the ladies. But naturalists have discovered that the topi, a midsize antelope, uses another surprising method: faking fear. When a topi spots a predator, it emits a snort that tells the potential attacker it has lost the element of surprise. During mating season, male topis play fast and loose with the truth. When a female in their territory is preparing to leave, the male will look in her direction and snort. The female will stop, assuming that danger lurks, and the male will grab one more mating opportunity. Conscious deception is rare among animals. Plovers are known to fake a broken wing to lead predators away from their nest, only to flap away to safety at the last moment. Knowing the truth and misrepresenting it requires what is known as a "theory of mind," the awareness that what you know is not the same as what everyone else knows. That's a very high-order insight. Animals that have it may be a lot smarter than we ever knew.

FROM TOP: BEVERLY JOUBERT/NATIONAL GEOGRAPHIC/GETTY IMAGES; PETER ARNOLD INC./ALAMY

Explosion of New Species in the South Pacific

Even after 4 billion years, nature never runs out of new product lines. That remarkable diversity was on display once again with the release of the results of a 2008 expedition to the Foja Mountains in New Guinea. The expedition ranged from the foothills of the mountains to peaks 7,200 feet high and was led by Conservation International as part of its Rapid Assessment Program (RAP), which oversees a sort of biological SWAT team that sweeps into a wilderness area and takes a fast census of its wildlife so that policymakers can have the information they need to protect new and endangered species.

Foja is a place well worth protecting. Over the course of millions of years, the Australian continental plate crept northward, plowing up the sea floor and causing it to compress into the New Guinea land mass. As the mountains slowly rose, the higher altitudes remained in a sort of permanent springtime—misty, cool, and stable—while the areas below grew much hotter. That wide range of environments made for a vast array of animals.

2.5 million

ACRES IN NEW GUINEA'S FOJA MOUNTAINS VIRTUALLY UNTOUCHED BY HUMANS

IDEOPSIS BUTTERFLY AND POTENTIAL NEW SPECIES
Just part of the new—and not yet fully described—menagerie

A bizarre little tree frog discovered by the expedition, for example, measures just under 2 inches and has a short, trunklike nose; when the male is calling, the nose inflates. A dwarf wallaby, a pintsize kin to the kangaroo, is no bigger than a small cat. A woolly rat—which weighs about 3.5 pounds and measures 18 inches from nose to tail—would scare the daylights out of anyone who encountered it on a city street late at night, yet is quite tame. Foja pigeons, common animals by most measures, have an elegant plumage in a mix of pearl gray, rust, and white.

What the expedition says about the biological exuberance of the Foja Mountains is almost secondary to what it says about the promise of projects like RAP. Federal and state agencies have developed similar programs, including one that is rolled out in the first stages of oil spills and other toxic events so that conservationists can know which species are being threatened and how best to protect them. If life has proved anything, it's that it can flourish almost anywhere. It's always easier to argue for conservation when we know which animals we're conserving and where they live.

FROM TOP: HENK VAN MASTRIGT; CONSERVATION INTERNATIONAL/STEPHEN RICHARDS

VET CARE *Dave Sharpe had trouble leaving his Yorktown, Va., home until Cheyenne helped ease the former airman's anxiety.*

COMBAT SOLDIERS SUFFER 10 TIMES AS MANY PSYCHOLOGICAL INJURIES AS PHYSICAL ONES. DOGS HELP TREAT **PTSD** SUFFERERS MERELY BY THEIR PRESENCE.

A Soldier's Best Friend? Sometimes a Dog

Staff Sgt. Brad Fasnacht was clearing mines on an Afghan road when an IED blast broke his spine and both ankles and put him in a two-week stupor that ended only when he woke up in the hospital. Although Army doctors and nurses have been able to get the 26-year-old walking again, they have been able to do little for his post-traumatic stress disorder (PTSD). For that, Sgt. Fasnacht has called in a specialist—Sapper, an Australian cattle dog mix.

With up to 400,000 cases of PTSD resulting from the Iraq and Afghanistan wars, finding new ways to deal with the condition is becoming crucial—and dogs are helping a lot. One study published as long ago as 1998 found psychiatric patients' anxiety dropped twice as much after spending 30 minutes with dogs as it did following other therapies. Multiple studies since have yielded similar results. Organizations with such playful names as Patriot Paws and Hounds4Heroes are now rushing to pair trained canines with soldiers like Fasnacht. The animals not only have a calming effect by their mere presence, but also help soldiers unburden themselves. When Afghanistan vet Dave Sharpe awoke from a war-related nightmare in a cold sweat, he found his dog, Cheyenne, staring at him. "What are you looking at?" he demanded. Cheyenne, undisturbed, simply barked, and Sharpe suddenly wrapped her in his arms and told her everything. "I have no idea why," he says, "but I felt completely at ease."

FROM LEFT: GILLIAN LAUB; MARTIN HARVEY/GETTY IMAGES; ROBERT DOWLING/CORBIS

Invasive Species Are Heeeeere!

The snake craze that caught on among American pet owners in the 1990s got out of control when python owners tired of the 20-foot creatures and began releasing them into the wild. The pythons have thrived there, particularly in the Everglades, where they pose a potential threat to humans and feed on native endangered species.

The pythons are hardly the only invasive species that is turning up where it shouldn't be. Asian carp, which catfish farmers imported to help clean algae from their ponds, have overrun the Mississippi River and are swimming toward the Great Lakes. Robben Island off South Africa—best known as the site of the prison that held Nelson Mandela—has become overrun by rabbits, brought in by Dutch settlers 300 years ago.

From zebra mussels (which hitched a ride to the northern Great Lakes aboard the hulls of U.S. merchant vessels from Europe) to starlings (released into New York's Central Park in the 1890s by a drug manufacturer who was trying to populate the park with every species of bird mentioned by Shakespeare), the story of invasive species has always been the same: Humans introduce them and then can't control them. None has ever been beaten, but the carp—the latest invaders—are at least facing resistance, with increasingly complex barriers that, engineers hope, will keep their range limited. The lesson, however, is a lot simpler: When you spot a species somewhere else in the world, leave it there—unless you want a whole lot of its descendants settling in for a very long stay.

$1.5 trillion

AMOUNT OF DAMAGE THAT INVASIVE SPECIES CAUSE EVERY YEAR, NEARLY 5% OF GLOBAL GDP

Engineering the Flu-Proof Chicken

Got a fever, sore throat, and all the other lovely symptoms of influenza? You can blame it on the birds. The main reservoir for influenza viruses is wild birds, which can pass the pathogen to domestic poultry, which potentially infect human beings. That's what has happened with the H5N1 avian flu, which has killed at least 306 people in the past seven years, not to mention hundreds of millions of chickens and other birds.

That could change with the recent news that a team of British scientists has genetically engineered chickens that, while still vulnerable to H5N1, don't seem to pass on the disease to other poultry. The trick was identifying a gene that could produce a piece of RNA that acts as a decoy to polymerase, an enzyme that is vital for viral replication. Rather than binding with the virus's genome, polymerase attaches itself to the decoy gene, preventing the virus from replicating. Transgenic chickens would have to be approved by regulators and accepted by consumers—no easy task. But if we can protect ourselves from the next flu pandemic by tweaking our birds, the benefits might be worth the Frankenstein factor.

Nonhumans Fighting Humanity's Battles

Animals have been part of our wars for a long time. Pigeons carry messages, horses carry soldiers, mules carry supplies, dogs sniff for bombs and casualties and alert sleeping soldiers when danger is approaching. Some unknowing creatures—including cows and monkeys—have even been dispatched into fields to serve as living (and often dying) land mine detectors. But in all those cases, the animals worked in a support capacity—getting killed, yes, but not doing any killing themselves.

That explains some of the outrage at a 2010 report in China's state-run *People's Daily* newspaper that the Afghan Taliban had begun training monkeys in areas along the Afghanistan-Pakistan border to attack occupying NATO forces. According to the story, the monkeys—mostly macaques and baboons—are hunted and captured when they are very young and trained to fire Kalashnikov rifles and trench mortars. They are also taught to recognize the uniforms of U.S. soldiers and allied forces. The report led to a lot of hyperventilating news reports about "monkey terrorists," and NATO sought to quiet the hysteria, denying any knowledge of the threat. But such cruel exploitation of animals that have no stake in humanity's wars is not without precedent. The CIA is said to have tried similar tactics during the Vietnam era. All this is causing some people to ask, So who's really the highest species?

WARRIOR SPIRIT *A blue monkey looking none too pleased*

ANOTHER VICTIM *Honeybees like this one are being wiped out by the millions.*

Searching for the Killer in the Die-Off of American Bees

It started in 2006. Scores of honeybees began dying for seemingly no reason, prompting scientists to come up with the term "colony collapse disorder" (CCD). According to the Department of Agriculture, reported bee-colony death rates in the U.S. were 29% in 2009, rising to 34% in 2010. Worse, it's not just honeybees anymore: A recent study by the University of Illinois suggests that the four main types of bumblebee populations have plummeted more than 90% in the past 20 years. About 130 crops—worth some $15 billion a year—depend on pollination from the honeybee alone in the U.S. It's scary to think what might happen to the world food supply if CCD can't be curbed.

It's still unclear what's behind the dying, but the Illinois paper provides some hope. The scientists are exploring the possible role of a parasite called *Nosema bombi,* which is common in European bumblebees but hasn't been fully investigated among North American populations. The study found that declining populations of American bumblebee species appear to be associated with high levels of parasite infection. Still, the authors write that the parasite could simply be more common in species that are declining—correlative, rather than causative. The uncertainty makes the deaths all the more eerie—and the need to find an answer all the more urgent.

Animal Suicide and Homicide

Something about SeaWorld trainer Dawn Brancheau's ponytail may have triggered the attack. That's what an official at the Orlando marine park told reporters the day after the 16-year veteran at SeaWorld was killed by Tilikum, a 12,000-pound killer whale. On Feb. 24, 2010, in the middle of a show, the 40-year-old trainer was standing at the edge of a tank when the 29-year-old animal leaped from the water, grabbed her by the ponytail, and dragged her underwater.

The tragedy raised all the usual questions about whether any wild animals—or at least big ones with big brains—should be kept in captivity. Nearly all zoo and park animals live better today than they did in the horror-show era of full-grown beasts in small metal cages. But many animal psychologists argue that the landscaping and enriched environments of modern zoos are as much for the benefit of human visitors as anything else. The dysfunctional behaviors on display at even the best zoos—from swaying giraffes to pacing big cats—illustrate the psychologists' point. "You can't enrich these environments," says Russ Rector, a former dolphin trainer and now a fierce opponent of keeping dolphins or whales in captivity.

The study of animal intelligence has advanced further and faster than the study of animal psychology, but if we're willing to acknowledge that, say, a dog can be sad, why can't a whale or a tiger be frustrated or bored or simply outraged at its confinement. Science can't answer that yet, but if zookeepers want to avoid more tragedies, they'd better try.

WHAT GOES ON IN THE MIND OF SO COMPLEX A CREATURE THAT CAUSES IT TO BECOME **SO FIERCE SO FAST** —AND IS THERE ANYTHING THAT CAN BE DONE TO PREVENT SUCH TRAGEDIES?

BETTER TIMES *Tilikum just months before killing his trainer*

IT JUST AIN'T SO...

Your Dog the Genius

We're hopelessly in love with our dogs—and why not? Their intelligence, empathy, and warmth make them irresistible. But many of those qualities may be illusions. Take the kiss you get from your dog when you come home. It looks like love, but it could be hunger. Wolves also lick one another's mouths, particularly when one wolf returns to the pack. They use their sense of taste and smell to see if the returnee has eaten some prey. If it did, the licking often prompts it to vomit up some of that kill for the other members of the pack to share. Our dogs may be giving us a similar inspection.

Then there's a dog's seeming self-awareness. We've all seen guilty dogs slinking away with lowered tails, but is that because they really understand that they've done something wrong? Alexandra Horowitz, a cognitive scientist at Barnard College, devised an experiment. First she observed how dogs behaved when they did something they weren't supposed to do and were scolded for it. Then she tricked the owners into believing the dogs had misbehaved when they hadn't. When the humans scolded the dogs, the dogs were just as likely to look guilty, even though they were innocent of misbehavior. What's at play here is not some inner sense of right and wrong but a learned ability to act submissive when an owner gets angry. "It's a white-flag response," Horowitz says.

There may also be less to a dog's ability to understand words than we think. Scientists at the Max Planck Institute in Leipzig, Germany, studied a border collie that knew the names of 200 objects. But the understanding probably stops with the single-word label. When it hears "Frisbee," a dog may think only, "Get the Frisbee." It will probably never recognize that Frisbee is a word that can be combined with other words to create sentences like "Run away from the Frisbee." None of this means dogs aren't worthy of our love and loyalty. They may simply be a little less worthy of our admiration.

Chemistry

CANCER WATCH ▪ CAT-CALMING SCENTS ▪ MEET THE NEWEST ELEMENT
BETTER BIOPLASTICS ▪ FUEL CELLS IN THE WORKPLACE ▪ THE CUDDLE CHEMICAL ▪ A NEW BANNED TOXIN
ARTIFICIAL PHOTOSYNTHESIS ▪ MICROBIAL HEROES OF THE OIL SPILL ▪ THE PIANO-PLAYING BRAIN

The Perils of Plastic

Industrial chemicals are everywhere—in our water bottles, toys, cosmetics—and some of them can do very nasty things in very tiny doses. But there are ways to fix the problem.

By Bryan Walsh

On the first Earth Day more than 40 years ago, the U.S. was a poisoned nation. Dense air pollution blanketed cities like Los Angeles, where smog alerts were a fact of life. Dangerous pesticides like DDT were still in use, and water pollution was rampant. Today air pollution is down significantly, the water is cleaner, and the pesticides have been eliminated or reformulated. But if the land is healing, Americans may be sickening. Since World War II, production of industrial chemicals has risen rapidly, and the U.S. generates or imports some 42 billion pounds of them a day, leaving Americans awash in a sea of synthetics. A recent biomonitoring survey by the Centers for Disease Control and Prevention (CDC) found traces of 212 environmental chemicals in Americans—including toxic metals like arsenic and cadmium, pesticides, flame retardants, and even perchlorate, an ingredient in rocket fuel.

"It's not the environment that's contaminated so much," says Dr. Bruce Lanphear, director of the Cincinnati Children's Environmental Health Center. "It's us." As scientists get better at detecting the chemicals in our bodies, they're discovering that even tiny quantities of toxins can have a potentially serious impact on our health—and our children's future. Chemicals like bisphenol A (BPA) and phthalates—key ingredients in modern plastics—may disrupt the delicate endocrine system, leading to developmental problems. A host of modern ills that have been rising unchecked for a generation—obesity, diabetes, attention-deficit/hyperactivity disorder—could have chemical connections.

If scientists were slow to arrive at that conclusion, Washington has been even slower. The Toxic Substances Control Act (TSCA), the 35-year-old vehicle for federal chemical

PHOTOGRAPHS BY JAMES DAY

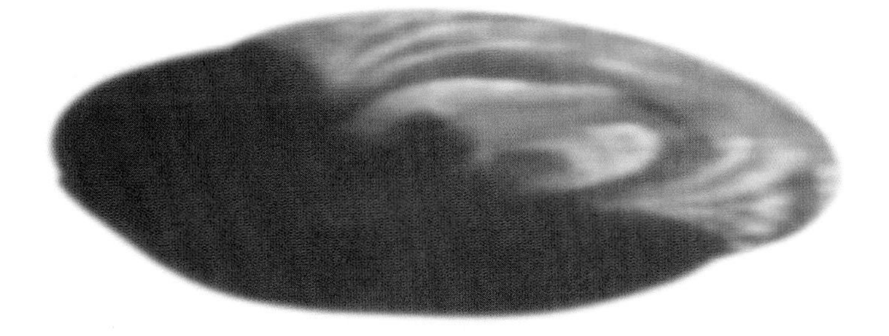

regulation, has generally been a failure. The burden of proving chemicals dangerous falls almost entirely on the government, while industry confidentiality rights built into the TSCA deny citizens and federal regulators critical access.

One of the reasons toxic chemicals have gotten a regulatory pass for so long is that they're often used in such tiny quantities. It was 16th-century Swiss physician Theophrastus Philippus Aureolus Bombastus von Hohenheim who established the dictum "The dose makes the poison," and that remains a bedrock principle of toxicology. But von Hohenheim never met BPA and the rest of the modern chemical stew, and he never dreamed of some of the techniques scientists would develop for diagnosing contamination. Biomonitoring systems can now detect exposure levels as small as one part per trillion, or about one-twentieth of a drop of water in an Olympic-size swimming pool. With that kind of resource at their disposal, scientists have realized that people are carrying far more chemicals than we'd thought.

At the same time, scientists learned that some toxins could harm at extremely low levels; the limit considered safe for lead, which can directly reduce IQ, has been lowered from 60 micrograms per deciliter of blood in 1970 to 10 micrograms today. Some chemicals, like BPA, may have strange effects even at very low doses. Invented in 1891, BPA has been used since the 1940s to harden plastics and make epoxy resin and is commonly found in the lining of food and beverage containers, among other products.

BPA performs its industrial job well, but that's not all it does. On the side, it also functions as a synthetic estrogen. Tiny amounts of hormones produce immense biological and behavioral changes, so it stands to reason that a chemical that mirrors a hormone might do the same, especially if a human being were exposed to it during critical periods of development, like the first trimester of gestation. That's exactly what dozens of scientists have found in animal studies, linking fetal BPA exposure in rodents to everything from mammary cancer to male genital defects and even neurobehavioral problems. Phthalates—chemicals used to soften plastics, found in products ranging from shower curtains to cosmetics—have similarly been shown to disrupt hormones in animals and have been linked to low sperm counts and other marks of feminization in male rodents.

While there are fewer studies of endocrine disrupters in humans, the ones that have been conducted have shown worrying associations. Higher levels of phthalates and other endocrine disrupters have been linked to earlier breast development in girls—a possible risk factor for breast cancer—and endocrine disrupters are a suspect in the rise in hypospadias, a correctable deformity of the urethra in boys.

Admittedly, the science around endocrine disrupters is far from settled. Studies show a correlation between phthalate exposure and developmental defects, but that doesn't mean the chemicals are causing the problems. Industry defenders point out that human exposure to BPA and phthalates is still well below safety levels set by the government and that health agencies around the world say the chemicals are safe for humans. And some peer-reviewed studies fail to show a positive connection between endocrine disrupters like BPA and health defects. "I think the research [on BPA] has been overhyped," says Richard Sharpe, an investigator at the Centre for Reproductive Biology at the Queen's Medical Research Institute in Edinburgh.

The scientific consensus, however, has been moving away from the idea that BPA is completely safe. In 2009, for example, the International Endocrine Society released a statement declaring that endocrine disrupters were a significant concern for public health and called for regulation to reduce human exposure. And even the FDA has changed its tune, expressing "some concern" over BPA. In the arid argot of a federal agency, that is dramatic stuff.

As scientists study the workings of toxins, policymakers must look closely at how industrial chemicals get approved for use and how that system can be tightened. If you want to market a new drug, you need to convince the FDA that it won't cause serious harm. But if you want to market a new chemical for use in a product—even one that will come into contact with children or pregnant women—it's up to the EPA to prove that it's unsafe, using whatever data are provided by the chemical company. The burden of proof has been turned upside down. Worse, if a chemical is one of the 62,000 already in use when the TSCA went into effect in 1976—and that includes BPA—chances are it was never tested by the government at all.

The result is a catch-22 for regulators and an information vacuum for consumers. The good news is that more than 30 years after the TSCA was signed, the pieces may finally be in place for much-needed retooling. Congress has been pushing various chemical-safety reform bills for the past two years, and even the chemical industry has admitted a need to reform the TSCA and is ready to negotiate. "Science has advanced a long way since the TSCA was adopted, and we recognize that more can be done to create a system that people have comfort and confidence in," says Cal Dooley, president and CEO of the American Chemistry Council.

Reform alone, though, won't defuse the basic debate over how much of an impact chemicals really are having on human health. Nearly everything we buy, sell, and use depends on chemicals, and the industry employs 803,000 Americans. Replacing some of the keystone ingredients of modern life would be challenging, not to mention costly. But challenging does not mean impossible, and costly does not mean unaffordable. With Americans' health and safety on the line, ignoring the problem is—or at least ought to be—unacceptable.

Codes of Concern *Recycling codes stamped on some plastics can help identify problematic chemicals.*

Type of polymer Polyvinyl chloride (PVC)

Uses Shampoo bottles, food packaging, medical equipment, shower curtains, pipes

Health concerns PVC may contain phthalates, which can pass from packaging into food, water, or cosmetics. The chemicals can also emit gas from curtains or pipes and be inhaled.

Type of polymer Polystyrene

Uses Styrofoam, some takeout containers, plastic utensils, insulation boards, plastic models

Health concerns It may leak the toxic chemical styrene. Polystyrene containers and cups are not biodegradable, posing a challenge for landfills.

10 CANCER PANEL RECOMMENDATIONS

1. Drink filtered tap water.

2. Store food and water in glass, stainless steel, or BPA- and phthalate-free containers.

3. Minimize children's and pregnant women's exposure to carcinogens and endocrine-disrupting chemicals.

4. Choose fruits and vegetables grown without pesticides or chemical fertilizers; wash all produce.

5. Choose free-range meat that has not been exposed to antibiotics or growth hormones.

6. Minimize consumption of processed, charred, or well-done meat.

7. Turn off lights and electrical devices when they're not in use.

8. Drive a fuel-efficient car; walk, bike, or use public transportation.

9. Check home radon levels.

10. Reduce radiation exposure from cellphones and medical tests; avoid UV overexposure.

Cancer Chemicals Everywhere?

Was it hype or prudence? In May 2010 the President's Cancer Panel published an alarming 240-page report on the risk of cancer from chemicals and other substances in the environment. "The true burden of environmentally induced cancer has been grossly underestimated," the report's authors concluded. "The American people—even before they are born—are bombarded continually with ... these dangerous exposures."

Scary stuff—and hard to ignore. The list of potential threats includes not just the ubiquitous BPA in plastics, but pesticides, exhaust from traffic, pharmaceuticals in the water supply, industrial chemicals, and radiation from medical tests, cellphones, and the sun. The authors of the report urge the government to increase research and regulation of these carcinogens, which pose "grievous harm," especially to children. More than 80,000 chemicals are on the U.S. market, of which only a few hundred have been proved safe, the authors note.

In large part, the panel's findings—chiefly that more research is needed—jibe with those of mainstream cancer researchers. But while the report highlights valid data, says Dr. Otis Brawley, chief medical officer of the American Cancer Society, its final conclusions may overreach. "There are environmental causes of cancer. We should not trivialize them, and we do need more research," he says, but the contention that the rate of environmentally caused cancers is "grossly underestimated" is not based in fact. "[The rate] very well may be higher [than the current estimate of 6% of all cancers], but the research has not been done to quantify that."

The take-home message: Minimize your exposure to toxins whenever you can, but don't be distracted from avoiding the known, major causes of cancers—smoking, obesity, alcohol, and sexually transmitted infections.

CLOCKWISE FROM TOP RIGHT: RANDY FARIS/CORBIS; GANDEE VASAN/GETTY IMAGES; FOTOG/TETRA IMAGES/CORBIS

Cat-Calming Scents

Even if you love cats, you have to admit that they can be a bit, well, high-strung. Cat owners with two or more hissing, fighting felines in the house have long looked for ways to get their tempers eased and their claws sheathed, and in recent years veterinarians have gone so far as prescribing very low doses of kitty Prozac. But there may be a better, more natural way to get the jungle out of your cat: pheromones.

Scientists have known for a half-century that animals communicate via pheromones, or scent chemicals, that do everything from trigger alarms to soothe offspring. Numerous studies in journals such as *Veterinary Record* and *Applied Animal Behaviour Science* now show that synthetic feline pheromones help reduce stress-related behaviors such as urine marking, vertical scratching, and aggression. Over the past year more than a million cat-owning households have used pheromone products—sold in sprays, collars, and even a plug-in air-freshener design. The chemicals have no effect on humans or other noncat species, and the effects can be seen almost immediately. The drawback: The products may need to be replaced after 30 days, so long-term use can get pricey. But calm in your home? Priceless.

RESEARCH INDICATES THAT SYNTHETIC FELINE PHEROMONES REALLY DO HAVE A CALMING EFFECT ON CATS—
LIKE A KITTY PROZAC
BUT WITHOUT THE PILL.

Meet the Newest Element

Lead, iron, and uranium are nothing compared to ununseptium, the temporary name for element 117, an extremely heavy combination of berkelium and calcium isotopes created in a particle accelerator in Dubna, Russia, in 2010. The new element existed for only the tiniest fraction of a second before vanishing again—and it must be independently created elsewhere before it earns a permanent spot on the periodic table of the elements. But the fact that it remained stable for even the fleeting instant that it did is promising. The heavier artificial elements get, the less stable they become, until they reach a point at which the curve turns back up and they begin to last longer and longer. Ununseptium is on the upward part of that arc, suggesting that what physicists call "islands of stability" may exist, at which the heaviest elements could last for months or years.

For now, the element will continue to be known by its unlovely name. But when it is formally accepted, it will be named after either the lab that created it or one of the scientists on the research team. Whatever the element is ultimately called, it's clear that the periodic table has yet to be fully set.

TAMARA SHOPSIN

Bioplastics Are Here—Sort Of

Regular, petroleum-based plastic doesn't biodegrade, which is why the water bottle you toss in a landfill today will still be around long after you're gone. One answer, as we've been told for decades now, is to switch to bioplastics—containers and packaging made of soybean or other vegetable byproducts that simply biodegrade. The truth, however, is a little more complicated than that.

Of the two most promising varieties of bioplastic, one—dubbed polylactic acid, or PLA—is clear in color and costs about 20% more to use than petroleum-based plastic. The other—called polyhydroxyalkanoate, or PHA—biodegrades more easily but is more than double the price of regular plastic. Both are made of fermented corn sugar, though some manufacturers are trying to make PLA using switchgrass, potatoes, and algae.

Price is a sticking point for industry and consumers. Another problem is that while the plastics do degrade, they do so only if you dispose of them properly—separating them from other trash so they can be sent off for composting. Mix bioplastic with ordinary plastic on trash-pickup day and not only will the plant-based waste not biodegrade, but it will also contaminate ordinary plastic during recycling. When you're trying to be green, good intentions must still be met by good execution.

BIOPLASTICS HAVE BEEN AROUND FOR DECADES. HENRY FORD MADE PARTS FOR THE MODEL T OUT OF CORN AND SOYBEAN OILS.

New Fuel Cells Mean Clean Power

Fuel cells are an old technology; they generate electricity within a cell through the reaction of a fuel and an oxidant. Essentially they're a kind of chemical battery. Car companies have long hoped to make hydrogen-fuel-cell-powered cars, but they've been limited by cost. That's beginning to change, thanks to a California startup called Bloom Energy. The company exploded onto the scene with the release of its Bloom Box, a system that uses solid oxide fuel cells in a box about half the size of a shipping container to create off-the-grid power. Bloom determined how to carry out the reaction at a relatively low temperature, making Bloom Boxes safe to use as a clean power source for corporate offices—which is how they're being put to work now by companies like Google and eBay—that want to reduce their carbon footprint. For those businesses at least, it's goodbye, dirty grid.

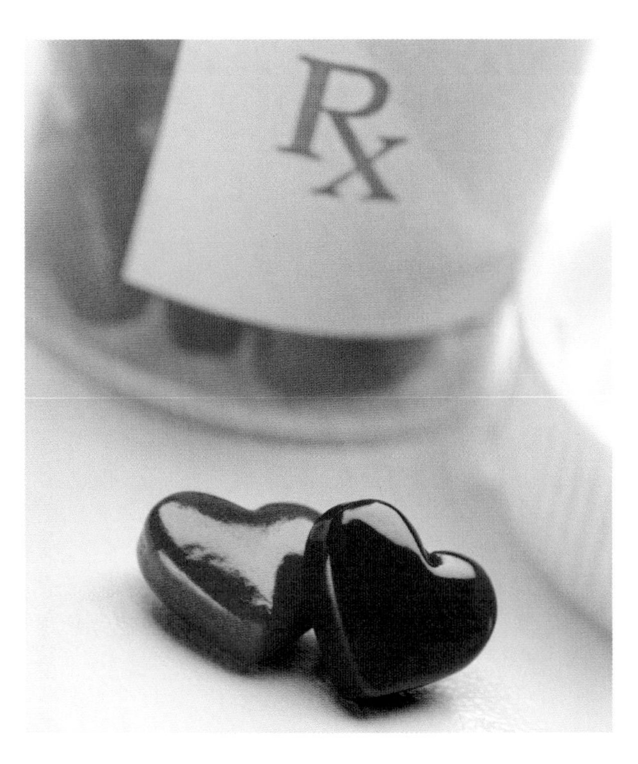

FROM TOP: ROBERT GALBRAITH/REUTERS/CORBIS; ROBERT GEORGE YOUNG/GETTY IMAGES

Get Your Dose of the Cuddlies

Next time you feel snuggly, thank the hormone oxytocin. Pumped out by the hypothalamus, oxytocin is known as the "cuddle chemical." It soars during labor and nursing and plays a major role in mother-baby bonding. New dads get a big slug of the stuff in their bloodstream too. Oxytocin is also released in men and women during sex.

Now it appears that you can get your oxytocin kick artificially. A team of German and British scientists recruited 48 men and divided them into two groups. Half received an aerosol dose of oxytocin and half got a placebo; all were then shown pictures of things like a crying child, a grieving man, and a girl hugging a cat. They were then asked to describe how deeply they felt the emotions associated with the pictures. On the whole, the men in the oxytocin group exhibited significantly higher empathy levels than those in the placebo group.

In theory, oxytocin doses could be used therapeutically on criminals or other people with antisocial tendencies. Such behavioral engineering, of course, would carry ethical implications. But still, the more we know about the chemistry of kindness, the more we can fathom what causes its absence.

IN ONE HOUR MORE ENERGY FROM THE SUN STRIKES THE EARTH THAN ALL THE ENERGY CONSUMED BY HUMANS IN A YEAR.

Artificial Photosynthesis: How Plants Could Make Fuel

As smart as human beings can be, nature almost always does it better—possibly because nature has had hundreds of millions of years to get it right. Take photosynthesis, for example. Plants with green leaves are able to capture the sun's energy and turn it into useful chemical fuel in a process that is much, much more efficient than our best photovoltaic solar panels.

That's why there are a number of scientists working on ways to create artificial photosynthesis. Daniel Nocera, an energy expert at the Massachusetts Institute of Technology, is pushing a form of artificial photosynthesis capable of creating electricity that would then be harnessed to produce hydrogen for use in fuel cells. That's only one way to harness photosynthesis, but already startups like Joule Biotechnologies are looking for ways to take it commercial.

Joule's system uses a solar concentrator that captures and focuses the sun's heat and light. This is then directed onto tanks containing a soup of brackish water, genetically engineered microorganisms, and nutrients. Carbon dioxide gas is stirred in as well, adding another critical photosynthetic ingredient. The microorganisms love all this and quickly set about doing what they were designed to do, which is to produce ethanol and other potentially valuable energy sources. That means biofuels without all the related problems that come with growing monocrops, such as corn, that gobble up land, distort food prices, and generate their own carbon footprint as they're converted to ethanol. Not bad—and it all comes from imitating nature.

Keeping Baby Clean—And Safe

Denmark's environmental ministry announced a ban on parabens in lotions and other products for children under 3. It is the first European country to ban the chemical preservative, which is suspected of being an endocrine disrupter and is used in a variety of soaps, deodorants, and other beauty products. No other European country has banned the chemical, but it was listed as a Category 1 substance (having evidence of endocrine-disrupting activity in at least one species in animal studies) by the European Commission in 2006. The U.S. Centers for Disease Control and Prevention is looking at the chemical too: One study showed that a type of paraben was present in 99.1% of urine samples of adults and children age 6 or older.

So is that a problem? Yes. While research has not conclusively proven that the chemical is a major threat to human health, it could be linked to irregularities in the reproductive system and hormone levels as well as to some kinds of birth defects. It also may play a role in breast cancer. Worse, children and babies, whose tissues are still developing, are at greatest risk. If you can't shop in Denmark, at least check labels when you buy.

FROM TOP: MAX OPPENHEIM/GETTY IMAGES; LAURENCE MONNERET

200 thousand

THE AMOUNT, IN TONS, OF METHANE GAS THAT WAS RELEASED INTO THE GULF ALONG WITH THE OIL

SICKLY SLICK *A toxic combination of fresh crude and chemical dispersant swirled on the surface of the Gulf of Mexico, nine miles from the site of the blown BP rig.*

The Microbes That Cleaned Up the BP Spill

As science experiments go, a massive oil spill isn't anyone's idea of a good field test. But while the more than 4 million barrels of oil spilled into the Gulf of Mexico after the sinking of BP's *Deepwater Horizon* rig mucked up coastlines and caused tens of billions of dollars' worth of damage, it also gave scientists an unprecedented chance to examine how a major water system would respond to all those hydrocarbons. The answer: surprisingly well. One reason: the microscopic bacteria in the gulf that digested much of the hydrocarbon load.

By weight, more methane was released from the BP wellhead than any other hydrocarbon. When scientists took water samples from the site while the oil was still flowing, they found significant amounts of methane and little evidence that bacteria were breaking it down. But when they came back months later, they found that the methane had largely disappeared. In its place were huge patches of water depleted of oxygen, evidence of action by bacteria, which use oxygen when consuming methane. Bacteria made similarly short work of the oil. Oxygen-depleted water is not a good thing, creating "dead zones" that can be inhospitable to life. But in the case of the methane, the depletion was only at a level of 40%, short of the two-thirds considered dangerous to fish. For all the human resources thrown at the spill, bacteria appear to have been the littlest heroes of a big disaster.

FROM TOP: CHRISTOPHER BERKEY/EPA/CORBIS; MARIO TAMA/GETTY IMAGES

The Piano-Playing Chemical

We may never know what made Mozart Mozart and most other people who sit down at the keyboard just, well, other people. But one answer might be GABA, a chemical messenger in the motor cortex of the brain that affects motor coordination and learning.

Investigators at the University of Oxford ran an imaginative experiment in which they took magnetic resonance images of subjects' brains both before and after a low level of electrical current was delivered through the scalp. The scientists already knew that such stimulation would cause a decrease of GABA in the motor cortex. They could thus get a reading of both the volunteers' baseline GABA levels and their so-called GABA responsiveness, or the ease with which the chemical adjusts itself in response to external triggers. On a later day the investigators asked all the subjects to try to learn a series of complex finger motions while their brains were scanned again.

The researchers found that the people who were more GABA-responsive learned the finger movements faster and more accurately. What's more, the motor cortex of their brains showed greater activation while they were practicing. Outside the lab—at the piano bench, say—a sensitive GABA response might help forge nerve connections that are essential for learning. The investigators also believe that their findings could provide clues to helping people recover from stroke or other brain trauma.

IT JUST AIN'T SO ...

Weight-Loss Hormone? Nope

Quick quiz: Does pregnancy cause weight loss or gain? It seems like a dumb question, but it's a test that the promoters of what are called hCG diets seem to have failed. Short for human chorionic gonadotrophin, hCG is the hormone secreted by the embryo that makes a pregnancy test positive. Since the 1950s some doctors have promoted hCG injections as the key to hunger-free weight loss—and now the diet is taking off on the web. This, despite 14 clinical trials showing that hCG has no effect on weight.

The hCG diet restricts caloric intake to 500 calories a day. That alone pretty much guarantees weight loss for anyone who sticks with it. But people who take a placebo instead of hCG while restricting calories do just as well as those who take the hormone—and taking the hormone doesn't increase the likelihood that people stay on the diet. Some doctors actually give injections of hCG, but many people take hCG pills, which are sold online—illegally, says the FDA. There's even less evidence for the effectiveness of the pills, and it's impossible to know if they even contain hCG.

There's also data to show that starvation-level diets—with or without hormones—can cause dramatic rebounds in weight in the long run, making maintaining healthy weight much more difficult. So if you want to find a diet that works, it's better to stick with the data and avoid hCG—unless you believe that a growing infant in the womb secretes hormones altruistically to avoid growing and to make its mom skinnier. Finally, consider this: Even the nausea of early pregnancy generally ends in weight gain, not loss.

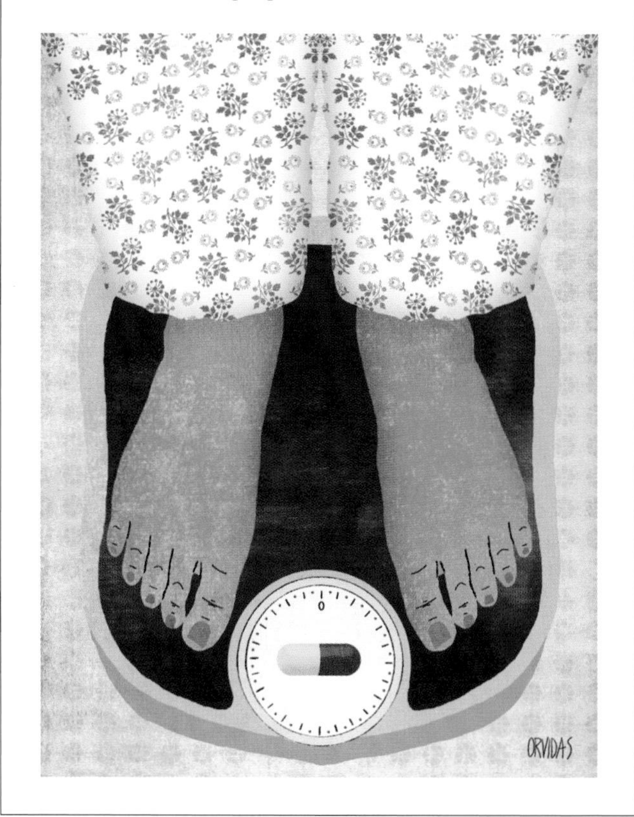

FROM LEFT: MATTHEW ANTROBUS/GETTY IMAGES; KEN ORVIDAS